Interactive GPU-based Visualization of Large Dynamic Particle Data

Synthesis Lectures on Visualization

Editors
Niklas Elmqvist, *University of Maryland*
David S. Ebert, *Purdue University*

Synthesis Lectures on Visualization publishes 50- to 100-page publications on topics pertaining to scientific visualization, information visualization, and visual analytics. Potential topics include, but are not limited to: scientific, information, and medical visualization; visual analytics, applications of visualization and analysis; mathematical foundations of visualization and analytics; interaction, cognition, and perception related to visualization and analytics; data integration, analysis, and visualization; new applications of visualization and analysis; knowledge discovery management and representation; systems, and evaluation; distributed and collaborative visualization and analysis.

A Guide to Visual Multi-Level Interface Design From Synthesis of Empirical Study Evidence
Heidi Lam and Tamara Munzner
2010

Interactive GPU-based Visualization of Large Dynamic Particle Data
Martin Falk, Sebastian Grottel, Michael Krone, and Guido Reina

ISBN: 978-3-031-01476-5   paperback
ISBN: 978-3-031-02604-1   ebook

DOI 10.1007/978-3-031-02604-1

A Publication in the Springer series
SYNTHESIS LECTURES ON VISUALIZATION

Lecture #8
Series Editors: Niklas Elmqvist, *Yahoo! Labs*
                David S. Ebert, *Purdue University*
Series ISSN
Print 2159-516X   Electronic 2159-5178

Interactive GPU-based Visualization of Large Dynamic Particle Data

Martin Falk
Linköping University, Sweden

Sebastian Grottel
Technische Universität Dresden, Germany

Michael Krone
University of Stuttgart, Germany

Guido Reina
University of Stuttgart, Germany

SYNTHESIS LECTURES ON VISUALIZATION #8

ABSTRACT

Prevalent types of data in scientific visualization are volumetric data, vector field data, and particle-based data. Particle data typically originates from measurements and simulations in various fields, such as life sciences or physics. The particles are often visualized directly, that is, by simple representants like spheres. Interactive rendering facilitates the exploration and visual analysis of the data. With increasing data set sizes in terms of particle numbers, interactive high-quality visualization is a challenging task. This is especially true for dynamic data or abstract representations that are based on the raw particle data.

This book covers direct particle visualization using simple glyphs as well as abstractions that are application-driven such as clustering and aggregation. It targets visualization researchers and developers who are interested in visualization techniques for large, dynamic particle-based data. Its explanations focus on GPU-accelerated algorithms for high-performance rendering and data processing that run in real-time on modern desktop hardware. Consequently, the implementation of said algorithms and the required data structures to make use of the capabilities of modern graphics APIs are discussed in detail. Furthermore, it covers GPU-accelerated methods for the generation of application-dependent abstract representations. This includes various representations commonly used in application areas such as structural biology, systems biology, thermodynamics, and astrophysics.

KEYWORDS

particles, visualization, GPU, molecules, rendering, visual analysis, object/image space methods, glyph rendering, atomistic visualization

Contents

Acknowledgments

This book is the result of more than a decade of active research. We have been supported by many people over the years, but we want to especially thank our common Ph.D. adviser, Professor Dr. Thomas Ertl, for support. We also want to thank our colleagues for their support and fruitful discussions as well as the undergraduate students supporting our work with theirs. Finally, our particular appreciation goes to the funding agencies that made this possible. The respective projects and agencies are, in no particular order:

- Landesstiftung Baden-Württemberg Project 688, "Massiv parallele molekulare Simulation und Visualisierung der Keimbildung in Mischungen für skalenübergreifende Modelle" (2004–2005)

- German Research Foundation (Deutsche Forschungsgemeinschaft, DFG) collaborative research center SFB 716 as subprojects D.3 and D.4 (2007–2018)

- German Research Foundation (Deutsche Forschungsgemeinschaft, DFG) Cluster of Excellence in Simulation Technology (EXC 310/1) (2008–2013)

- Centre Systems Biology in Stuttgart, Germany (2007–2010)

- Excellence Center at Linköping and Lund in Information Technology (ELLIIT) (2013–2015)

- Swedish e-Science Research Centre (SeRC) (2013–present)

- Federal Ministry of Education and Research (BMBF) Project No. 01IS14014, Scalable Data Services And Solutions (ScaDS, 2014–present)

- European Social Fund (ESF) Project No. 100098171, Visual and Interactive Cyber-physical Systems Control and Integration (VICCI, 2012–2014)

Martin Falk, Sebastian Grottel, Michael Krone, and Guido Reina
September 2016

Figure Credits

Figure 1.1	From: S. Grottel, P. Beck, C. Müller, G. Reina, J. Roth, H.-R. Trebin, and T. Ertl. Visualization of Electrostatic Dipoles in Molecular Dynamics of Metal Oxides. *IEEE Transactions on Visualization and Computer Graphics*, 18(12):2061–2068, 2012a. Copyright © 2012 IEEE. Used with permission.
Figure 1.2	Courtesy of Mathieu Le Muzic.
Figure 2.3	From: A. Knoll, Y. Hijazi, A. Kensler, M. Schott, C. Hansen, and H. Hagen. Fast Ray Tracing of Arbitrary Implicit Surfaces with Interval and Affine Arithmetic. *Computer Graphics Forum*, 28(1):26–40, 2009. Copyright © 2009 John Wiley & Sons, Inc. Used with permission.
Figure 3.3	From: S. Grottel, G. Reina, and T. Ertl. Optimized Data Transfer for Time-dependent, GPU-based Glyphs. In *IEEE Pacific Visualization Symposium (PacificVis 2009)*, pages 65–72, 2009a. Copyright © 2009 IEEE. Used with permission.
Figure 5.7	From: M. Le Muzic, J. Parulek, A. Stavrum, and I. Viola. Illustrative visualization of molecular reactions using omniscient intelligence and passive agents. *Computer Graphics Forum*, 33(3):141–150, 2014. ISSN 1467-8659. Copyright © 2014 John Wiley & Sons, Inc. Used with permission.
Figure 5.8	Courtesy of Mathieu Le Muzic.
Figure 5.9	Courtesy of Mathieu Le Muzic.
Figure 5.10	Based on: M. Hopf and T. Ertl. Hierarchical splatting of scattered data. In *IEEE Visualization 2003*, 2003.
Figure 5.11	From: M. Hopf and T. Ertl. Hierarchical splatting of scattered data. In *IEEE Visualization 2003*, 2003. Copyright © 2003 IEEE. Used with permission.
Figure 5.12	From: M. Hopf and T. Ertl. Hierarchical splatting of scattered data. In *IEEE Visualization 2003*, 2003. Copyright © 2003 IEEE. Used with permission.

Figure 7.4   From: S. Grottel, G. Reina, C. Dachsbacher, and T. Ertl. Coherent Culling and Shading for Large Molecular Dynamics Visualization. *Computer Graphics Forum,* 29(3):953–962, 2010a. Copyright © 2010 John Wiley & Sons, Inc. Used with permission.

Figure 8.6   From: K. Bidmon, S. Grottel, F. Bös, J. Pleiss, and T. Ertl. Visual Abstractions of Solvent Pathlines near Protein Cavities. *Computer Graphics Forum,* 27(3):935–942, 2008. Copyright © 2008 John Wiley & Sons, Inc. Used with permission.

Figure 8.7   From: S. Grottel, G. Reina, J. Vrabec, and T. Ertl. Visual Verification and Analysis of Cluster Detection for Molecular Dynamics. *IEEE Transactions on Visualization and Computer Graphics,* 13(6):1624–1631, 2007. ISSN 1077-2626. Copyright © 2007 IEEE. Used with permission.

Figure 8.8   From: T. Ertl, M. Krone, S. Kesselheim, K. Scharnowski, G. Reina, and C. Holm. Visual Analysis for Space-Time Aggregation of Biomolecular Simulations. *Faraday Discussions,* 169:167–178, 2014. ISSN 1364-5498. Copyright © 2014 Royal Society of Chemistry. Used with permission.

CHAPTER 1

Introduction

Particle-based data has been an important subject of research for many years now. In its simplest form, such data consist of just a spatial coordinate. Each of those infinitesimal points can however be extended via additional attributes, hinting at some fitting representing geometry, color, or physical attributes like radius, velocity, mass, charge, or the like. A fundamental property of such data is the lack of any topological information. Each particle is an object of its own and does not require additional data to be represented meaningfully and individually. Relevant sources for this kind of data are mainly computational in nature, like molecular dynamics simulations and granular media simulations. Depending on the application area, the available attributes, and the research questions asked, a multitude of metaphors has been explored to depict such data.

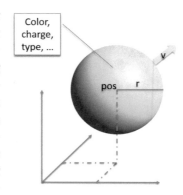

1.1 SCOPE OF THIS LECTURE

This lecture will cover the interactive visualization of large, dynamic particle data consisting of discrete entities with their individual semantics. Given the computational power available today, the resulting data sets can grow quite large, currently in the order of 10^4 to 10^{12} [Eckhardt et al., 2013]. Simulating 10^{12} particles requires entire HPC systems (thousands of nodes) to be effective, so data of this magnitude is still uncommon at the time of writing. To further delineate the kind of data this lecture will focus on, it needs to be distinguished from what is nowadays called "big data." The most important additional property of such data over what we call "large" is that it cannot be effectively post-processed or, more specifically, visualized using a single workstation. This definition intentionally scales with the evolution of hardware and thus needs to be considered in context, but from a practical point of view, "large" is constant, describing feasibility for a current researchers' desktop workstation, so to say. Conversely, the approaches presented here enable visualization on a single machine and fulfill the additional requirements that a) only negligible preprocessing time is required, if any, thus enabling a domain scientist to directly inspect the latest simulation results; b) static data or data from at least a single time step must fit in host memory, so visualization is in-core with respect to the static case; c) the data must be streamed to the GPU since it is usually dynamic and it will not necessarily fit into GPU memory entirely; and d) consistent interactivity of the approach so the user can explore data and visualization parameters freely.

Since more elaborate visualization algorithms require more compute time, a tradeoff between data set size vs. computational complexity can be necessary to ensure interactive visualization. Thus, it is important to keep in mind that "large" data is relative to the visualization algorithm, not only to the hardware capabilities. Especially the interactivity requirement suggests the use of modern, programmable GPUs, where moderate implementation effort already results in highly interactive visualizations. Solutions that solely make use of the CPU exist, but here interactivity requires a much more involved implementation even for the baseline visualization [Wald et al., 2015]. In this case, large data sets are visualized using machines that have a considerably higher cost than the systems targeted by our approaches, a four-socket XEON system with 3TB RAM vs. what is basically a commercial off-the-shelf (COTS) gaming PC (consider that the list price of just a single XEON 8890v3 CPU as in the paper at the time of writing costs more than 7,000 USD, which exceeds the cost of the whole gaming PC by a factor of more than two). There are also approaches that solve the GPU memory problem at the cost of preprocessing and by reducing data accuracy somewhat, like the approach presented by Reichl et al. [2014], but the general idea will always be more severely memory limited or accuracy limited than when relying on the more ample host memory.

Application areas that make day-to-day use of simulations with the aforementioned properties include:

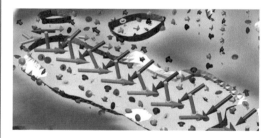

Figure 1.1: Direct and aggregated visualization of dipole behavior in crack propagation. From [Grottel et al., 2012a].

Physical sciences Many subfields in physics exploit particle simulations to study the behavior of systems under certain conditions. Thermodynamics and process engineering investigate phase interfaces and transitions, specifically evaporation and nucleation, as well as wetting or coating of surfaces, mixing of substances, and so on. Computer physics and material science investigate the inner structure of solids, i.e., crystal lattices and granular media, and how they behave under external influences like ablation, stress, rupture, and crack propagation.

Life sciences Biochemistry investigates the components of living organisms ranging from simplified scenarios like the interaction of proteins and solvent or DNA behavior over viruses to the processes in whole cells, like signal transduction. The combination of simulation and visualization has also been dubbed a *computational microscope* [Dror et al., 2012, Lee et al., 2009], since molecular simulation has nowadays advanced to a point where they can partially substitute wet lab experiments. With such a tool, the observation of small-scale systems is less costly than resorting to real-world experiments and expensive magnification equipment. Another significant advantage is that hazardous substances or environments are cheaper to handle and cannot pose any physical threat to the scientists involved.

In general, visualization can also be employed to check simulations for plausibility via arbitrary navigation in time and space. As a practical example, a domain scientist used to visually inspecting simulations is able to tell whether a simulation has been performed at some temperature that is completely wrong just from the movement and interaction of particles over time. Visualization can also serve for dissemination of results or education of students by explicitly showing the effect of underlying laws and principles.

The data set sizes in practical use vary significantly depending on application area and use case. The number of attributes per particle is usually low enough to guarantee that a data set stays roughly in the same order of magnitude, however data can become arbitrarily large if long-term effects or rare events need to be observed. Thousands of time steps are commonplace despite the simulations already emitting only every $Nth(N > 1)$ time step, so data

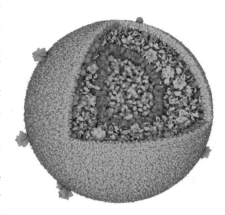

Figure 1.2: Level-of-detail rendering of a HI virus as generated by cellVIEW. Courtesy of Mathieu Le Muzic.

scales linearly with the number of time steps (e.g., by at least three orders of magnitude for a thousand time steps). To give an idea about the common size of a single step, one can start with simple proteins with thousands of atoms and thus only megabytes in size. Data is considerably larger in the physical sciences, where millions of particles are simulated and a time step can already require gigabyte(s) of memory. The upper end is represented by life sciences, where a HI virus including blood plasma was represented using 15 billion atoms [Le Muzic et al., 2015] which would require in the order of 200 GB just for storing an uncompressed time step, and cosmological simulations, like the DarkSky data set, where a single time step consists of one trillion particles and requires more than 30 TB [Skillman et al., 2014].

The described scope can be employed to derive a few simple requirements for particle-based visualizations. These lead to the concrete approaches that will be presented in the following chapters. First off, the resulting visualization must be intelligible, at best without any prior training. This means the metaphors used should be closely related to known approaches from textbooks or some other status quo the user can relate to. With few to no additions to the textbook prototypes, the result can be directly used for a broader public, meaning students of the respective topic or even for general dissemination. Interactivity is also key, since arbitrary exploration is oftentimes a central feature that distinguishes state-of-the-art visualization from the status quo found in commonly used visualization tools. Last, scalability to complex systems should be given. This can be achieved by technical measures plus abstractions that allow for better performance and easier, possibly hierarchical, exploration of extremely large data.

1.1.1   BASE-LINE RENDERING

Independent from the actual research question, the availability of a direct visualization is important throughout the analysis process and the evolution of problem-specific visualization approaches. This base-line rendering is essential for the user to trust in more complex visualizations, since it can always serve for comparison purposes and verification of more abstract visualization results. Exploratory analysis fosters unbiased, novel findings, as every abstraction cements assumptions by way of the transformed visualization. Its fidelity is also required for the scientist to check simulation results for plausibility even at a very small scale (visual debugging): a common result of a researcher being confronted with a visualization of his/her simulations for the very first time is the realization that something is unexpected or wrong. The first case is a good starting point for a scientific publication, while the second one at least helps improving the simulation model or implementation (in turn potentially leading back to case one).

The base-line rendering should be a neutral starting point when communicating with any application scientist. It should correspond to the unabridged raw data set, usually presented via a common-ground metaphor. For molecular dynamics the calotte (also called space-filling or CPK) model is used, resulting in hard spheres for every available atom. Here, the radius of the sphere equals the atomic radius of the corresponding chemical element. Granular media can have different particle representations, for example as instances of a pre-defined set of available geometries of varying complexity [Grottel et al., 2010b]. In this context the lecture will cover different visual metaphors and the according GPU-based rendering methods to achieve interactivity even for large numbers of particles (see Chapter 3). Additionally, concise data representations will be discussed in Chapter 5, easing storage but more importantly improving performance when transferring data to the GPU.

The direct representation of particles presents some issues as well. For example, this strategy cannot scale indefinitely: if we consider an arbitrary output device with a physical resolution of $w \times h$ pixels, it stands to reason that no more than $w * h$ particles in the projected image plane can be discerned. The resulting image will of course mostly present noise since all particles are just single colored pixels in this limit case, so the real limit is considerably lower. The actual data set will conversely be bigger than the product $w * h$ if the domain is $\Re^3$. Part of the complexity in depth is lost due to projection and occlusion. The analogy is still fitting as resolution is an effective cap for data set size if a faithful overview is to be given via base-line visualization. Interactive camera adjustment can of course partially remedy this issue, since it allows users to zoom into regions of interest and get a detailed view of interesting regions in the data set. The problem to identify these interesting regions in the data set, however, is still present since the base-line rendering does not always easily show structures formed by the particles in very large data, especially if they are on different scales.

1.1.2   HIGHER-LEVEL STRUCTURE

Practical experience shows us that we can render about 10^8 particles interactively without taking special measures, but even with high-resolution output the result is difficult to understand in its higher-level structure as this is hidden by the high-frequency details when just using local lighting and a simple shading model. One solution to this problem is the introduction of additional visual cues via global lighting effects or transparency, as detailed in Chapter 7. On the other hand, even though the data does not contain a topology, there is often a higher-level functional structure available. The solution is usually application-specific so the structures have hugely differing sizes and complexity, starting from droplets in thermodynamics, over amino acids and proteins in biochemistry to whole cells in systems biology. These structures can, for example, be summarized by their interface to the environment. Thus, metaphors result in application-specific visualizations like ellipsoids for droplets or molecular surfaces in different configurations and their respective semantics (see Chapter 8).

1.2   RELATED TOPICS BEYOND THE SCOPE

This lecture explicitly does not go into detail about two variants of particle-related visualizations. First, particle systems as they are employed for special effects, for example representing sparks, fire, or smoke. The visual metaphor here abstracts from the single particle to generate larger-scale visual effects geared toward a believable visual imitation of reality that does not necessarily require physico-chemical correctness, but to conserve the artistic appearance [Feldman et al., 2003, Reeves, 1983]. Second, particles representing a sampling of continuous functions. This category can itself be divided in two specific areas: point set surfaces and smoothed particle hydrodynamics (SPH).

The first subset, point set surfaces, describe surfaces which require explicit or implicit reconstruction from a limited number of discrete samples of an object, e.g., obtained by laser scanning. This has its own caveats as any holes can either be an artifact stemming from limited resolution or be intentional. The surfaces of such point sets are rendered with point-based methods like splatting or surface elements (surfels) [Pfister et al., 2000]. The GPU lends itself to efficiently render such surfaces in high quality [Botsch and Kobbelt, 2003]. The most recent point cloud rendering approaches benefit from the latest GPU features and utilize vertex buffer objects and bindless textures for improved performance [Goswami et al., 2013]. For more details, the reader is referred to the comprehensive overview over the whole process from acquisition over data handling to rendering as edited by Gross and Pfister [2007]. The idea of point-based surfaces has also been applied to the field of visualization, e.g., the rendering of stream surfaces in 3D flows [Schafhitzel et al., 2007].

Smoothed particle hydrodynamics (SPH) on the other hand allow the reconstruction of a density-based data set from discrete samples by way of a so-called *smoothing kernel*. This method is often employed for fluid simulations, and respective surface rendering approaches have been presented. Examples include the work by Zhang et al. [2008] and Reichl et al. [2014], utilizing the GPU for both simulation and rendering.

Another prominent application for SPH is cosmology, employing this method for the simulation of dark matter distribution following the big bang. Several large-scale visualization techniques have been presented in the last decade, for example by Hopf and Ertl [2003] or Fraedrich et al. [2009].

Despite the differences, this whole body of work still represents a good source of information, especially regarding some of the technical aspects. Data handling and quantization are examined in detail and will still help today if the data can be condensed sufficiently to make a whole data set fit on the GPU, taking data transfer problems out of the equation entirely. Rendering optimizations can also be relevant, however, since much of the prominent work is about a decade old, recent additions to GPU features and API capabilities are not covered, and much has changed especially regarding memory management (bindless access, persistent mapping, raw shader storage, etc.). For more details on optimizations and GPU features, we refer the interested reader to the Radeon Rays SDK using OpenCL[1] and the OptiX Ray Tracing Engine relying on CUDA.[2]

---

[1]Radeon Rays SDK | AMD, Inc.: http://gpuopen.com/gaming-product/radeon-rays/ (last accessed 08/19/2016).
[2]OptiX™| NVIDIA Corp.: https://developer.nvidia.com/optix (last accessed 08/19/2016).

CHAPTER 2

History

The scenarios and requirements sketched out in the introduction have been a widespread application case in visualization for at least 30 years now [Max, 1983]. As such, the widely accepted solutions presented in this lecture shall be briefly put into historical context. Very well-known and early examples for particle visualization were all applied to molecular dynamics, for example VMD [Humphrey et al., 1996], PyMOL [Schrödinger, LLC, 2015], Chimera [Pettersen et al., 2004], and AtomEye [Li, 2003]. These tools offer comprehensive functionality that goes far beyond visualization and is specifically geared toward the application. However, they also are "legacy" in the sense of having been designed before the widespread availability of programmable GPU. Having grown over time, they support hardware of diverse capabilities, but are not organized from the ground up for optimal performance when run on a capable GPU. This concerns memory management as well as the traditional polygon-based rendering. VMD, for example, also supports some of the techniques that will be described further on, but as the focus is on generality, this is just an option and not the base strategy. Looking at polygon-based rendering especially in the context of molecular dynamics brings us to the observation that the basic element, a sphere, is rather costly in tessellated form.

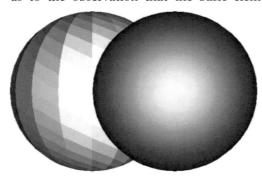

Memory is not a concern if instancing is employed. However, a spherical silhouette is only obtained using a considerable number of triangles—tens to hundreds, depending on the screen-space footprint (see Figure 2.1). The resulting load on the geometry stages of modern GPUs is thus quite high. This can be improved upon using multiple LoDs or dynamic tessellation from some basic representation like a tetrahedron, but the effort is better spent otherwise, as we will show in Chapter 3. Shading does not present a problem, as it can be performed per fragment and, thus, has optimal quality.

Figure 2.1: Tessellated sphere geometry with per-triangle shading and per-fragment shading.

Subsequently, a better-scaling approach for large numbers of atoms was presented by Bajaj et al. [2004]. TexMol employs an imposter in form of a depth and normal texture so each atom needs only a bounding geometry—a quad—that is correctly scaled and placed as a billboard. Using the depth replace feature in the fragment shader, the initially flat billboard is modified to even allow for depth-correct intersections of spheres.

Local shading can be performed based on the normal texture. The result is much faster than the geometry-based approach and offers nearly comparable quality: intersections are more precise than with discretized spheres, however shading presents slight banding artifacts from the limited precision of the employed normal texture. The billboard-based approach also does not allow for perspective-correct rendering and as such reduces the available depth cues.

This kind of strategy was an excellent fit at the time, but this is not the case anymore, as it causes high memory pressure due to the large number of texel fetches and has negligible arithmetic intensity. Modern GPUs, however, are designed to actually require lots of computations to hide the relatively high latency of their memory interface. Considering a Geforce FX 5900 Ultra (from 2004) and a Geforce 980 Ti (from 2015), the respective memory bandwidth has increased from about 27 GB/s to about 337 GB/s (12.4-fold). The respective compute power, however, has increased from about 15 GFlops/s to more than 6,000 GFlops/s (400-fold). Consequently, modern rendering techniques should expend many arithmetic operations if a texel fetch can thus be avoided.

Figure 2.2: A protein data set rendered using TexMol [Bajaj et al., 2004]. The inset shows banding artifacts due to the limited precision of the normal texture.

These changes in GPU technology led to the current state-of-the-art. A desirable technique would compute a correct surface together with a normal and the resulting shading from as terse a representation as possible for each visible fragment. One classical approach is ray casting, i.e., ray tracing using only primary rays. Implementing a purely image-order ray caster requires a spatial data structure to reduce the number of hit tests against the whole scene. The implementation is a bit more intricate, but performs very well for very large data (see Chapter 5). A hybrid image/object-order approach can take advantage of the graphics pipeline to render billboards which tightly fit the desired geometry, similar to the approach in TexMol. However, the geometry itself is only represented implicitly, so very few parameters suffice to describe basic solids (position, orientation, dimensions). These parameters can be directly attached to the proxy geometry and can thus be stream-processed together in the graphics pipeline. As a result, the geometry load for the GPU is low, same as the memory pressure. In return, high arithmetic intensity is generated by the ray-object intersection tests, depending on the complexity of the represented geometry. Implicit functions up the order of four (i.e., quartics) can easily be solved explicitly, while higher-order surfaces require iterative approaches. Performance behaves also very beneficially: the more particles there are, the smaller their screen-space footprint becomes. The resulting geometry load is not increased significantly because of the simplicity of the billboards, while the fragment shader load is balanced by the relative size. The result fits modern GPU architecture quite well, so several approaches based on this concept have been presented over the years. A few examples will be shown before proceeding to the detailed explanation of the fundamental hybrid approach in the next chapter.

The starting point was marked by Gumhold [2003]. He presented depth-correct implicit ellipsoids which are ray cast in the fragment shader. The concept was then refined by Klein and Ertl [2004]. Here, the ray is first transformed into the respective parameter space of each glyph, thereby simplifying the problem to intersecting a unit sphere. In the same year, Toledo and Lévy [2004] presented general quadrics to be rendered on the GPU.

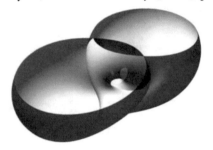

Figure 2.3: Arbitrary implicit shape using interval arithmetic. From [Knoll et al., 2009].

Then, application-specific glyphs were developed, for example a dipole representation Reina and Ertl [2005] as well as very costly iterative hyperstreamtubes for DTI rendering Reina et al. [2006]. Sigg et al. [2006b] wanted to reduce computational cost and thus improved the bounding geometry to fit implicit quadrics more tightly. More generalized solutions were also devised: Loop and Blinn [2006] showed an approach to generate algebraic surfaces by way of Bézier tetrahedra, while Knoll et al. [2009] even allowed arbitrary implicits by employing interval arithmetic (Figure 2.3).

CHAPTER 3

GPU-based Glyph Ray Casting

Point-based visualization seems like a natural choice for particle-based data sets. In fact, this approach is an optimal representation of the data, as it is true to the original simulation results. The basic idea is that each element in the data set, in the case of data sets from MD simulations atoms or rigid molecules, is depicted by a compact graphical element—a *glyph*—representing the principal attributes of the data element. There are several different approaches for implementing glyph-based visualizations. In this chapter we cover GPU-based ray casting of glyphs.

The core idea of ray tracing is to trace the light energy as it is distributed throughout the scene. In a reverse fashion, light is tracked from the observer in the specific direction through a single pixel, through the scene, until it hits an object. The required viewing ray through the image pixel is easily computed from the observer position, viewing direction, and camera parameters like aperture angle, viewport resolution, and so on. This ray is then intersected with the closest object's surface. From this surface hit point, the light is further traced back, for testing light source visibility (shadow ray) and through reflection or refraction rays (other secondary rays) through and to other objects. This approach is well suited for glyph-based particle visualizations, as the glyphs' surfaces can be described mathematically in a straightforward way.

Ray casting (also called ray marching) simplifies this technique by only performing local lighting calculations (usually Blinn-Phong illumination [Blinn, 1977]). This approach allows for precise per-fragment evaluation and shading of the surface of a glyph, thus creating a perfect apparent image for each glyph anew, at the exact resolution required. Full ray tracing is usually not required by visualizations, as the advance effects in light transport, e.g., refractions and reflection, are not helping in understanding the data, and might even distract the user and create wrong impressions. Additionally, since ray casting does not require secondary light rays, it can be efficiently implemented on the GPU's fragment shader stage, resulting in high frame rates and not requiring additional data structures. Therefore, for a base-line visualization we opt for simple glyph ray casting, and extend this approach for more advanced visualizations (see Chapter 5).

The fundamental principles of glyph ray casting were initially presented by Gumhold [2003], who described how to efficiently evaluate a ray-surface intersection with programmable graphics hardware. The surface of the glyph was described by a quadratic equation, allowing for a general solution of many surfaces of this class, e.g., sphere, ellipsoid, cylinder, and cone. Klein and Ertl [2004] presented a similar approach, focusing on ellipsoids to aid their visualization goals, but utilizing the point primitive of OpenGL to optimize the required data transfer between GPU and main memory as well as evaluating the intersection in parameter space. This way, the ray was

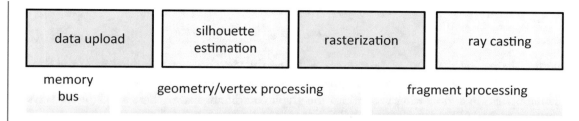

Figure 3.1: Simplified process of GPU-based ray casting: The blue stages are part of the rendering hardware, while the green stages are operations of the ray casting, usually performed in programmable shaders. First the size of the glyph's silhouette is estimated in vertex processing to span an image-space quad. Then, for all fragments of this quad the ray casting is performed.

deformed instead of the unit sphere, simplifying calculations considerably. Figure 3.1 illustrates the principal steps of the general technique. These resulting glyph images are perspective-correct and can intersect each other by emitting a per-fragment correct depth and applying the built-in OpenGL depth test. The fragment shader stage is used to perform the viewing ray surface intersection (cf. Section 3.1). In the vertex shader stage the graphical primitive—we could also call it billboard—is fit to an approximation of the final silhouette of the glyph in image space (cf. Section 3.2). This generates the fragments required for the actual ray casting of the glyph.

3.1   FRAGMENT-BASED RAY CASTING

The probably simplest glyph for particles, and also the most commonly used one, is the sphere:

$$|\boldsymbol{x} - \boldsymbol{p}| = r \ , \tag{3.1}$$

with $\boldsymbol{p}$ being the position and r the radius of the sphere. Thus, $\boldsymbol{x}$ describes all points on the surface of the sphere. To simplify the required ray casting calculations we introduce a glyph coordinate space, in which the sphere is placed at the origin, eliminating $\boldsymbol{p}$:

$$|\boldsymbol{x}| = r \ . \tag{3.2}$$

This coordinate system transformation is described as a matrix, similar to the well-known model matrix. This new coordinate system seems exaggerated effort, and it surely is not obvious for spheres. For more complex glyph types, however, this matrix can also include rotation and scaling. A corresponding example for cylinder glyphs is given below. Using the glyph coordinate system to bring these glyphs into a default orientation is very convenient for ray casting. Ensuring, e.g., a cylinder's base axis to be one of the world coordinate system major axes not only simplifies the whole ray casting into a 2D problem, but implements the required projection by simply removing one of the coordinates.

The equation of a viewing ray is then given by:

$$c' + \lambda v' = x \, , \qquad (3.3)$$

with c' being the position of the camera, v' being the viewing vector derived from the current fragment, and λ being the ray casting parameter. Both, c' and v' are obtained by transforming camera position and viewing vector into the glyph coordinate space. There are different ways of setting up this viewing ray in an actual fragment shader. Listing 3.1 shows a generic solution by back projecting the fragment coordinate `gl_FragCoord` into glyph space. Apart from the inverse `ModelViewProjection` matrix this requires the pixel resolution of the frame buffer. `viewAttr` stores $2 * width^{-1}$ and $2 * height^{-1}$ in the z and w component. Please note that in this shader code the glyph space is only a translated object space, i.e., the case for sphere glyphs. For more complex glyphs line 9 would be replaced by the matrix vector multiplication to bring `coord` into glyph space (similar to line 7).

Listing 3.1: GLSL fragment shader segment performing view ray setup

```
1  // transform fragment coordinates from window coords to view coords.
2  vec4 coord = gl_FragCoord
3    * vec4(viewAttr.z, viewAttr.w, 2.0, 0.0)
4    + vec4(-1.0, -1.0, -1.0, 1.0);
5
6  // transform fragment coordinates from view coords to object coords...
7  coord = gl_ModelViewProjectionMatrixInverse * coord;
8  coord /= coord.w;
9  coord -= objPos; // ... and to glyph space
10
11 // calc the viewing ray
12 vec3 ray = normalize(coord.xyz - camPos.xyz);
```

Figure 3.2: Zoomed-in view; visual artifacts result from numeric errors.

While the direction vector v is a given for all computations, the starting point of the view ray can be arbitrarily chosen along the ray itself. For the sake of simplicity we use c, the position of the camera. This is intuitive and easy to compute. However, this can become a numeric problem. Consider ray casting a very large image or a virtual camera with a very small aperture angle. As a result the v vectors of neighboring pixels only differ slightly. In fact the differences in direction after normalization easily are of the same order as the smallest values representable by floats. Each v thus contains a numeric error. This error is scaled by λ, cf. Equation (3.3), resulting in inaccuracies proportional to the camera distance (thus, larger scenes behave worse). Figure 3.2 shows an example rendering containing the resulting visual artifacts. Due to the error in the viewing ray the glyph surface is evaluated at the wrong positions. To resolve this issue, either the error in v needs to be decreased, e.g., by using double precision variables, or the value of λ needs to be decreased. While the first solution can have significant performance impact, the second one can be implemented by choosing c to be close to the glyph center. During silhouette approximation, cf. Subsection 3.2, the vertex shader will compute a depth based on the glyph center point. The resulting initial fragment position can directly be used as a good alternative for c.

Combining Equations (3.3) and (3.2) results in the quadratic equation for the first surface hit point:

$$|c + \lambda v| = r \, ,$$

$$(c_x + \lambda v_x)^2 + (c_y + \lambda v_y)^2 + (c_z + \lambda v_z)^2 = r^2 \, ,$$

$$\lambda^2(v_x^2 + v_y^2 + v_z^2) + 2\lambda(c_x v_x + c_y v_y + c_z v_z) + c_x^2 + c_y^2 + c_z^2 - r^2 = 0 \, ,$$

$$\lambda^2(v \cdot v) + 2\lambda(c \cdot v) + (c \cdot c) - r^2 = 0 \, ,$$

$$\lambda_{1,2} = \frac{-2(c \cdot v) \pm \sqrt{(2(c \cdot v))^2 - 4(v \cdot v)(c \cdot c - r^2)}}{2(v \cdot v)} \, ,$$

$$\lambda_{near} = \frac{-(c \cdot v) - \sqrt{(c \cdot v)^2 - (v \cdot v)(c \cdot c - r^2)}}{(v \cdot v)} \, . \tag{3.4}$$

Note that only the λ for the surface point closest to the camera is required, thus $\pm$ simplifies to $-$. Formulated in terms of vector operations, only three dot products are necessary, which nicely match GPU shader capabilities. Optimizing the view ray setup by normalizing v further simplifies to:

$$\lambda_{near} = -(c \cdot v) - \sqrt{(c \cdot v)^2 - (c \cdot c) + r^2} \, . \tag{3.5}$$

The corresponding GLSL fragment shader segment is given in Listing 3.2. Matching the set up from Listing 3.1, the glyphs size is explicitly use in line 6. sqRad is the square of the radius of the sphere.

Listing 3.2: GLSL fragment shader segment performing sphere ray casting

```
1  // calculate the geometry-ray-intersection
2  float d1 = dot(coord.xyz, ray);
3  // projected length of the cam-sphere-vector onto the ray
4  float d2s = d1 * d1 - dot(coord.xyz, coord.xyz);
5  // off axis of cam-sphere-vector and ray
6  float radicand = d2s + sqRad;
7  // square of difference of projected length and lambda
8  if (radicand < 0.0) discard;
9
10 float sqrtRadicand = sqrt(radicand);
11 float lambda = - d1 - sqrtRadicand;
12 vec3 sphereintersection = lambda * ray + coord.xyz;
```

Figure 3.3: Examples for compound glyphs; left: arrow glyphs constructed from cylinders and cones; right: glyph for molecules of the cooling agent heptafluoropropane, constructed from cylinders and spheres. From [Grottel et al., 2009a].

Line 8 discards the fragment if the sphere is not hit at all. Since the found hit point `sphereintersection` is in glyph space, the surface normal vector can be calculated trivially.

Note that all occurrences of $(v \cdot v)$ from Equation 3.4 are missing. In our viewing ray set up, the direction vector v is normalized, and therefore the result is equal to one. This approach works well for all shapes which can be compactly described, e.g., using a closed-form representation or distance-field representation. If the hit point of viewing ray with the shape cannot efficiently be calculated analytically, a numeric hit-point search is always possible. The usual approach would be an initial approximation, followed by an iterative linear search, either using fixed or dynamic step size. Finally the hit point can be refined using a recursive bisection search. A concrete example of such a numeric approach, however, is not in the scope for this chapter.

One further practical aspect is the composition of rigid glyphs. Figure 3.3 shows two examples for such glyphs. Generic arrow glyphs, constructed from a cylinder and a cone, are obviously useful for a wide range of applications, including flow or motion visualization, or simply annotations. In many simulations, individual particles represent complex but rigid elements, like molecules without inner degrees of freedom. Water molecules are a prominent example, but even larger cases exist, like the molecules shown on the right side of Figure 3.3.

All such compound glyphs can be defined as a fixed group of simple primitive glyphs. Usually the glyph is only placed, rotated, and scaled as a whole, based on the particle's attributes. Therefore, it makes sense to render the glyphs in a single pass as well. For a detailed study on compound glyphs, see Grottel et al. [2009a].

The fundamental idea is to ray cast all the primitive shapes in one shader. The final result simply is the valid hit point closest to the viewer, i.e., with the smallest ray casting parameter λ. The ray casting can be performed in the individual glyph spaces of the primitive shapes independently, if the corresponding ray casting vectors v are not normalized independently. Note that consequently the simplification regarding $(v \cdot v)$ mentioned above is not valid anymore. So the shader code shown in Listing 3.2 needs to be extended matching Equation 3.4. If, for example, the view ray is constructed in world space, transferred into the different glyph spaces, and not re-normalized afterward, then the resulting λ values will all be compatible. The smallest lambda can then directly be used on the original view ray in world space.

Furthermore, there is some potential for optimizations. For example, the order in which the hit test calculations are performed can be based on a quick sorting in combination with some trivial rejects. Consider a glyph constructed from several spheres. If the hit point found from intersection with the first sphere has a larger distance to the center of the remaining spheres than their radii, further hit test calculations can be skipped. Such an optimization was presented by Reina and Ertl [2005] for glyphs of twin-center Lennard-Jones molecules with dipole charges.

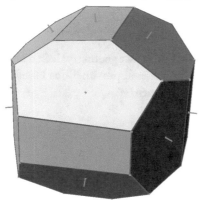

Another example for optimization is presented by Grottel et al. [2010b] for grain glyphs, cf. Figure 3.4. In this application, grain glyphs are defined by the union of multiple half-spaces. Technically, they are given by vectors—direction and length—from the common glyph center. Ray casting these grains is trivial, as only plane ray intersections need to be computed. To find the closest valid hit point, however, most hit points need to be rejected. The first rejection strategy is to remove or invalidate all hit points of the planes hit from the wrong side. This test is performed by $v \times h$ where h is the vector of the half space plane. As h is also normal to the half space plane, the sign of the result specifies which side of the plane the viewing ray hits, i.e., the plane is hit from the inside if positive. This is the analog to back-face culling but evaluated based on normal vectors instead of using a winding rule. Then, from all remaining (front-side) hit points, the final point needs to be selected. As the construction rule is based on union operations, each half space reduces the size of the final glyph. Therefore, the one correct hit point is the one farthest away from the viewer, i.e., with the largest λ.

Figure 3.4: Grain glyph rendering. The grain is defined as a union of multiple half-space definitions, in this case 18 planes.

3.2   SILHOUETTE APPROXIMATION

Most calculations of our rendering approach are performed in the fragment shader, i.e., the ray casting and surface shading. However, some support geometry needs to be constructed to first generate the fragments in question. Rendering an image-space filling quad for each particle is obviously a bad idea. Each fragment will carry out most of the ray casting computations even if the fragment is then discarded. Thus, the support geometry needs to be tight fitting. The precise geometry of the image of the glyphs, however, is expensive to compute and difficult to generate. Therefore, the fundamental concept is to cheaply approximate the silhouette of the final glyph in the vertex shader, and to create the supporting geometry based on this approximation.

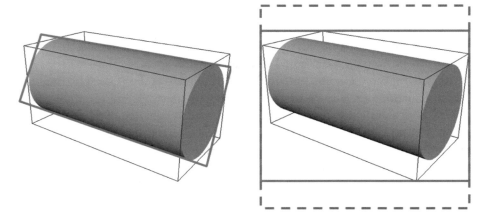

Figure 3.5:   Silhouette approximations for cylinders, assuming a quad support geometry; left: a near-optimal approximation, but potentially expensive to compute; right: cheap approximations based on the projection of corners of the glyph-space bounding box into image space.

3.2.1   BOUNDING BOXES

The most generic approach to the problem of approximation is to use simple bounding volumes, the easiest ones being bounding boxes. Following the core idea mentioned above, the corner points of the bounding box simply need to be projected into image space. The convex hull of these points then approximates the image of the projected glyph. Figure 3.5 visualizes this concept. On the left side a near-optimal approximation is shown. Such a support quad can be computed if the glyph shape is simple enough [Grottel, 2012]. The following Subsection 3.2.2 contains a corresponding example for spheres.

The right side of Figure 3.5 shows a more generic solution: The core idea is to approximate the convex hull in image space by a bounding rectangle (brown rectangle). This obviously increases the number of fragments which will be computed and discarded. However, the computation of this image space quad is trivial, both for the programmer and for the GPU. The thus faster vertex shader stage usually even outweighs the additional fragment shader invocations. This trade-off can be further optimized for each concrete application scenario. As a rule of thumb, the number of vertices increases linearly with the number of particles. The number of fragments, however, does not, because when more particles are visualized, the individual particles become smaller and cover fewer fragments each. Thus, in terms of scaling, vertex shader optimizations are more important than fragment shaders.

Nevertheless, the first step of this silhouette approximation is to project the bounding box corner points into image space. This is performed using classical computer graphics algebra and, e.g., a model-view-projection matrix M. Note that M and all vectors use four-dimensional homogenous coordinates (indicated by a tilde):

$$v = \begin{pmatrix} x \\ y \\ z \end{pmatrix} \Rightarrow \begin{pmatrix} x \\ y \\ z \\ 1 \end{pmatrix} \qquad \tilde{v} = \begin{pmatrix} x \\ y \\ z \\ w \end{pmatrix} \Rightarrow \begin{pmatrix} x/w \\ y/w \\ z/w \end{pmatrix}, \qquad (3.6)$$

where the homogenous component w is 1 for points and 0 for vectors.

Naïvely all eight coordinates of the bounding box would be generated by matrix-vector multiplications. However, in homogenous coordinates the matrix multiplication is a linear mapping. Therefore, it is sufficient to only transfer four linearly independent positions into image space. The other four positions can be reconstructed before performing the perspective division.

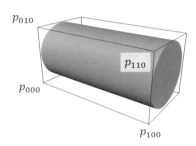

The following details the example of a cylinder glyph with radius r and length l. In glyph space the rotational axis of the cylinder coincides with the x axis. The eight corners of the glyph space bounding box are then given by:

$$\boldsymbol{p}_i = \begin{pmatrix} \pm l/2 \\ \pm r \\ \pm r \end{pmatrix}.$$

Figure 3.6: Glyph-space bounding box of a cylinder, with corner point $\boldsymbol{p}_i$.

Figure 3.6 shows the cylinder bounding box with four corners explicitly marked. Postponing the division by w allows one to work with linear combinations of projected vectors. In glyph space, it is obvious that

$$\boldsymbol{p}_{110} = \boldsymbol{p}_{100} + \boldsymbol{p}_{010} - \boldsymbol{p}_{000}.$$

All four positions can simply be rewritten with homogenous vectors:

$$\tilde{p}_{110} = \tilde{p}_{100} + \tilde{p}_{010} - \tilde{p}_{000},$$

all extended with $w = 1$. As stated above, the multiplication with the model-view-projection matrix M is distributive:

$$M\tilde{p}_{110} = M(\tilde{p}_{100} + \tilde{p}_{010} - \tilde{p}_{000}),$$

$$M\tilde{p}_{110} = M\tilde{p}_{100} + M\tilde{p}_{010} - M\tilde{p}_{000}.$$

Thus, any number of points for a bounding geometry can be projected into image space with only four matrix-vector multiplications. The following Listing 3.3 shows the corresponding excerpt of a vertex shader.

Listing 3.3: GLSL vertex shader segment projecting the bounding box for the image space approximation

```
1   mat4 mvp; // model-view-projection matrix
2   vec2 rad = vec2(RAD, -RAD); // signed radii for fast access
3   // vec3 POS = glyph position
4   ...
5   // four matrix multiplications for the projected coordinate system
6   vec4 p000 = mvp * vec4(POS + r2.yyy, 1.0);
7   vec4 p100 = mvp * vec4(POS + r2.xyy, 1.0);
8   vec4 p010 = mvp * vec4(POS + r2.yxy, 1.0);
9   vec4 p001 = mvp * vec4(POS + r2.yyx, 1.0);
10  // x is not required in this case
11  vec4 y = p010 - p000;
12  vec4 z = p001 - p000;
13  // compute the four remaining corners
14  vec4 p110 = p100 + y;
15  vec4 p101 = p100 + z;
16  vec4 p011 = p010 + z;
17  vec4 p111 = p110 + z;
18  // now perform all perspective projections aka div-by-w
19  p000 /= p000.w;
20  p100 /= p100.w;
21  p010 /= p010.w;
22  p110 /= p110.w;
23  p001 /= p001.w;
24  p101 /= p101.w;
25  p011 /= p011.w;
26  p111 /= p111.w;
```

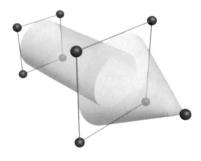

Figure 3.7: Arrow glyph with several glyph-space points defining a convex hull; note the point at the tip of the arrow and the different radii.

After this projection the image-space bounding rectangle can be easily computed using min and max operations. This approach is not restricted to glyph-space bounding boxes, but to any convex bounding constructs. Figure 3.7 shows the corresponding example for an arrow glyph. A classical bounding box would obviously not be a very good approximation, due to the different thickness along the glyphs symmetry axis. Instead, a convex bounding volume of a pyramid and a pyramid frustum is defined by the nine points shown in the figure. But, again, four matrix-vector multiplications suffice.

The vector computations after the matrix-vector multiplications are cheap, but not completely ignorable. There is thus a limit to how fine-grained a glyph-space bounding geometry should be. Following the argumentation of large data

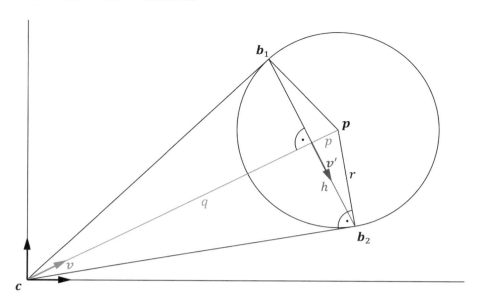

Figure 3.8: Evaluation of the horizontal image-space extent of a sphere at position p with radius r; The image-space extent as seen from camera c is given by the points b_1 and b_2.

again: when showing many particles, the individual glyphs become small. Thus, if the, e.g., 3×3 fragments result from a bounding box or from a bounding truncated octahdron (just an arbitrary example) it does not make any difference in the visual result. Conversely, however, the complex bounding geometry is more expensive in the vertex stage.

3.2.2   SPHERES

For specific shapes for the glyphs, like closed quadratic surfaces, e.g., spheres or cylinders, the image-space silhouette can be computed analytically. This subsection presents the corresponding calculations for spheres. The silhouette must be calculated as forward projection: finding the image-space bounding rectangle based on the surface parameters. In theory all surface points are projected from object space into image space and the bounding rectangle enclosing all points is evaluated. Obviously the goal is to only transfer and process on the GPU the four points on the surface which will contribute to the final result, i.e., the points which will have the minimum-/maximum x/y-coordinate after projection.

Spheres are simply defined by a position and radius in object space. Finding the four contributing points on the surface of the sphere is analytically possible and has been shown in Sigg et al. [2006a] and Reina [2008] before. The image-space bounding rectangle corresponds to an asymmetric pyramidal frustum, thus given by four planes with common attributes: all planes go through the camera position and all touch the sphere as tangent planes. Additionally the normal vectors of the planes are reduced in their degrees of freedom as they have additional constraints from the shape of the frustum: for the plane defining the horizontal extent in image-space the y-component of the normal vector needs to be zero. Thus these calculations can be reduced to a problem in 2D: to the x-z-plane for the horizontal extents and to the y-z-plane for the vertical extents. Figure 3.8 depicts the principle of calculating this extent for a sphere at position p with radius r. The solution is given by the two contact points b_1 and b_2 of the sphere and the tangent planes through the camera position c. Considering the triangle $c\,p\,b_2$ the values of the distances p, q, and h are given following the Pythagorean theorem:

$$p = \frac{r^2}{|p - c|} \qquad q = |p - c| - p \qquad h = \sqrt{pq} \, . \tag{3.7}$$

Given $v = p - c/|p - c|$ being the normalized vector from the camera position to the position of the sphere and v' forming a orthonormal basis with v in this 2D plane the requested points b_1 and b_2 can easily be calculated:

$$b_{1,2} = p - pv \pm hv' \, . \tag{3.8}$$

Note that v' results trivially from v by switching components and adding negating one component, because this computation takes place in a 2D subspace. Thus the calculation of the four points in object space which will span the image-space bounding rectangle is computationally cheap. The four points can then be projected into image-space and directly mapped to a bounding rectangle. A similar approach for cylinders is, e.g., presented in Grottel [2012]. This sphere silhouette approximation is implemented by the shader code excerpt in Listing 3.4.

Listing 3.4: GLSL vertex shader segment performing bounding box projection and image space approximation

```
1   vec2 d, d_inv, p, q, h; // use vec2 to compute vertical and horizontal
2                           // subspaces at the same time
3   d.x = length(cam_proj_yz); // projected camera vector
4   d.y = length(cam_proj_xz);
5   d_inv = vec2(1.0) / d;
6   // triangle lengths
7   p = squarRad * dd;
8   q = d - p;
9   h = sqrt(p * q);
10  p *= d_inv;
11  h *= d_inv;
12  // scale vectors, because we need them multiple times
13  cam_proj_yz *= p.x;    // cf. v
14  cam_proj_i_yz *= h.x; // cf. v' in figure 3.8
15  cam_proj_xz *= p.y;    // cf. v
16  cam_proj_i_xz *= h.y; // cf. v'
```

```
17   // project and test
18   vec3 objTangPos = objPos.xyz + cam_proj_yz + cam_proj_i_yz;
19   vec4 projPos = gl_ModelViewProjectionMatrix * vec4(objTangPos, 1.0);
20   projPos /= projPos.w;
21   // ... and the other three points accordingly
22   // min-max selection based on the constructed subspaces
```

Note that the shader code performs this approximation in the two subspaces, y-z and x-z, at the same time, by using vec2 for scalar values. Since modern GPUs use scalar processors there is no performance benefit. This just allows one to write the core calculations as single lines instead of duplicating these lines. porjPos is the final image-space position of the tangent point at the sphere's silhouette we were aiming for. The *min-max selection* mentioned in line 22 will result in the smallest window-axes aligned rectangle enclosing the whole image of the sphere glyph.

3.3   GEOMETRY GENERATION

The previous subsections defined ways to calculate the size of the required image-space support geometry. What is missing is the technical approach to actually create this geometry. For this we have several options, ranging from explicit bounding boxes, over using quads, using OpenGL point primitives, to using modern OpenGL constructs like the geometry shader or the tessellation shader.

3.3.1   BOUNDING BOX

An extremely simple but valid approach to generate the required fragments is to just use the glyph-space bounding box itself. The whole silhouette approximation (cf. Sec. 3.2) can then be omitted. The bounding geometry is explicitly drawn, e.g., using GL_TRIANGLES. Similar to explicit, geometry-based graphics, the vertex shader simply converts the vertex positions into the image-space coordinate system. The fragments created will naturally cover the image of the glyph. Therefore, a ray casting fragment shader will work just fine.

For very simple glyphs and especially when working with large data, this approach has several drawbacks. The pure vertex data load is increased by a factor of eight for bounding boxes. Of course, a bounding tetrahedron could also be used for simple shapes. All the same, the increased data load has a performance impact on the data transfer when streaming particle data to the GPU, as well as on the vertex processing. On the one hand, consider the large data case. The individual particles will cover only a few pixels, while the large number of vertex shader invocations still remains. Therefore, it is desirable to drastically reduce the number of vertices per particle, even if this results in a more complex vertex shader.

On the other hand, for data scenarios in which the individual geometry elements will retain a certain image-space size, or if the ray casting setup is rather complex, using an explicit bounding box as support geometry can be the best solution. Chapter 5 will present such scenarios.

3.3.2   QUAD PRIMITIVE

The obvious first data reduction from an explicit bounding box is to use an image-space bounding "quad." The vertex load is reduced from ×8 to ×4, but more importantly the number of triangles is reduced from 12 (at most six after back-face culling) down to two. In return, however, the vertex shader now needs to be more complex. Note that the GL_QUADS primitives are deprecated, but we can use triangle strips or triangle fans as replacement. Connectivity of triangles (cf. glPrimitiveRestartIndex) and the winding rule (cf. glFrontFace), e.g., if the constructed triangles are front facing, are simple to overcome hurdles.

The vertex shader evaluates the image-space silhouette as discussed above in Section 3.2. Then each vertex shader invocation selects the 2D image-space coordinate for one of the quads corners. A simple way of selecting the corners is to use gl_VertexID (cf. Listing 3.5).

Listing 3.5: GLSL vertex shader segment selecting one quad corner based on the gl_VertexID

```
1  vec3 minPos, maxPos; // the image-space results from
2                       // the silhouette approximation
3  int selX = gl_VertexID % 4; // 4 corners for each quad
4  int selY = selX / 2;        // 0, 1 -> 0 and 2, 3 -> 1
5  selX %= 2;                  // 0, 2 -> 0 and 1, 3 -> 1
6  gl_Position = vec4(
7    (selX == 0) ? minPos.x : maxPos.x,
8    (selY == 0) ? minPos.y : maxPos.y,
9    minPos.z,
10   1.0);
```

The input values minPos and maxPos need to be valid 3D vectors, i.e., the division by w must have already been performed. There are alternatives to the ternary conditional operators in lines 7 and 8. For example the mix function could be used (cf. Listing 3.6).

Listing 3.6: GLSL vertex shader selecting a quad corner using a boolean vector

```
1  vec3 minPos, maxPos; // the image-space results from
2                       // the silhouette approximation
3  bvec2 selVal = bvec2((gl_VertexID % 2) == 1,
4    (gl_VertexID % 4 / 2) == 1);
5  gl_Position = vec4(
6    mix(minPos.xy, maxPos.xy, selVal),
7    minPos.z,
8    1.0);
```

This approach reduces the number of graphics primitives and thus the load on the vertex processing and rasterization engine. But, compared to the simple bounding box approach, the vertex shader complexity is drastically increased, especially since four vertex shader invocations basically calculate almost the same values. Detailed performance analysis is presented in Chapter 4.

3.3.3 POINT PRIMITIVE

Obviously, instead of performing almost the same calculations in four vertices for a single particle, one would want to only use one vertex for each particle. The OpenGL GL_POINT graphics primitive provides a viable solution. Points, in OpenGL, are single-vertex primitives, which are rasterized into an image-space square. The size of the square is specified in pixels. The point size can be specified in host code using the function glPointSize or in the vertex shader using the built-in output variable gl_PointSize. The latter needs to be enabled using GL_PROGRAM_POINT_SIZE. For compact glyphs in particle visualization, this seems like a perfect fit. Note that OpenGL points support rendering as anti-aliased circular sprites, enabled by GL_POINT_SMOOTH. This option needs to be disabled for our approach. We only need the fragments to be generated and will handle the shading ourselves.

Particles are uploaded as single vertices using GL_POINT. Thus the data transfer is near optimal, not considering additional spatial data structures and compression. The single vertex shader instance computes the image space extents and parametrizes the point accordingly (cf. Listing 3.7).

Listing 3.7: GLSL vertex shader parametrizing a GL_POINT primitive

```
1  vec3 minPos, maxPos; // the image-space results from
2                       // the silhouette approximation
3  gl_Position = vec4((minPos + maxPos) * 0.5, 0.0, 1.0);
4  // from normalized device coordinates into pixel coordinates
5  maxPos = (maxPos - minPos) * 0.5 * winSize;
6  gl_PointSize = max(maxPos.x, maxPos.y) + 0.5; // round up
```

One minor flaw is the fact that the points are always rasterized as image-space squares. Images of glyphs might be elongated and thus using a square increases the number of wasted fragments. Consistent with the scenario of large data and small particle images, this, however, is not really an issue.

There are also some severe problems with this approach. First off, the maximum size for points is limited to a hardware vendor specific constant. This is oftentimes something as small as 64×64 pixels. Zooming into particle data results in clipping artifacts as soon as particles exceed this image-space size. Many modern graphics cards, however, have a rather large maximum size, something in the order of the display resolution. For example, on Windows 10, x64, an Nvidia GeForce GTX 960, reports a maximum size of 2047 pixels (which will, however, vary depending on driver or hardware revision, at least we observed this in the past). So, depending on your hardware, this might not be an issue after all.

A second possible source of visual problems is the clipping behavior. Per default, points are clipped at the one position, i.e., their center positions. As such, particles, of which almost half would be visible, are still removed as a whole. This issue is not that problematic in desktop environments, as the particle removal only occurs at the border of the window. In a tiled-display environment, however, this issue is severe, as half a particle will go missing within the image at inner display boundaries. At least on Nvidia GPUs this problem can be addressed by setting the point size in host code sufficiently large. Then points are clipped based on their square extent as defined by this size, instead of based only on their center position. The final number of fragments generated, however, is still controlled by the vertex shader output. Thus, setting the point size in host code does not impact the rendering performance. This behavior, however, does not follow the OpenGL specification. You thus cannot rely on it and it might change any time. A more sophisticated solution is to increase the image size, i.e., viewport, framebuffer, and view frustum, but only perform fragment operations on the size of the originally targeted image size, i.e., setting a corresponding scissor region and using the scissor test.

3.3.4   GPU-SIDE GENERATED QUAD PRIMITIVE

Modern OpenGL provides additional means to get the best of both approaches: minimal vertex data upload and actual quad geometry for clipping and rasterization. The two solutions utilize the geometry shader and the tessellation unit to generate quad geometry from single vertices directly on the GPU. The vertex load remains minimal, as only a single vertex shader invocation per particle is required. But in both cases additional rendering pipeline stages will be enabled, resulting in some overhead, e.g., synchronization points in the data stream.

Using the geometry shader seems like an obvious choice, as this shader type can convert between different types of graphical primitives. Converting points into triangles, e.g., a triangle strip of two triangles forming a quad, is a typical task for these shaders. The geometry shader stage was infamous for being slow, for one due to being too generic and for two due to the serial vertex data emission in one shader invocation. This has been shown, for example, by Grottel et al. [2009b]. However, on modern GPUs the geometry shader has proven itself to be fast and to be a viable option in many applications.

When using a geometry shader in this scenario, the distribution of work between it and the preceeding vertex shader is important. Basically, as for the GL_POINT based approach above, the vertex shader still performs all calculations which would be the same for all four vertices of the image-space quad. The geometry shader simply picks the right coordinates and emits the final vertices. Listing 3.8 shows the whole geometry shader.

Listing 3.8: GLSL geometry shader to create an image-space quad after the glyph's silhouette was estimated by the vertex shader

```
1  #version 430
2  layout(points) in;
3  layout(triangle_strip, max_vertices = 4) out;
4  flat in vec3 vertColor[];
5  flat in vec3 vertMinPos[];
6  flat in vec2 vertMaxPos[];
7  layout(location = 0) flat out vec3 geoColor;
8  layout(location = 1) flat out vec4 geoPosRad;
9  void main() {
```

```
10    geoPosRad = gl_in[0].gl_Position;
11    geoColor = vertColor[0];
12    gl_Position = vec4(vertMinPos[0].xyz, 1.0);
13    EmitVertex();
14    gl_Position = vec4(vertMaxPos[0].x, vertMinPos[0].yz, 1.0);
15    EmitVertex();
16    gl_Position = vec4(vertMinPos[0].x, vertMaxPos[0].y,
17      vertMinPos[0].z, 1.0);
18    EmitVertex();
19    gl_Position = vec4(vertMaxPos[0].xy, vertMinPos[0].z, 1.0);
20    EmitVertex();
21    EndPrimitive();
22  }
```

Note the linear emission of vertices, one of the reasons the geometry shader is often claimed to be slow. However, since the shader itself does not perform any computations but just a data selection, the performance impact of the linear output can be assumed to negligible. Also, the number of output vertices is constant four, meaning the GPU's data scheduling should be able to optimize the pipeline.

An alternative approach to keep full parallelization is to use the new programmable tessellation unit. The fundamental idea of the tessellation unit is to create graphical primitives (quads or triangles) approximating a surface patch defined by a number of input vertices. This includes input patch definitions with only a single vertex. The tessellator generates these graphical primitives for the target patch, i.e., the connectivity information, according to the parameters set by the tessellation control shader. The tessellation evaluation shader specifies the actual positions of the fine-grained vertices. The vertex shader runs before the control shader and can, again, calculate everything for the whole particle. The control shader is hard-coded to generate a single quad (cf. Listing 3.9). The evaluation shader selects the right corner coordinates (cf. Listing 3.10), similar to the quad approach (cf. Section 3.3.2).

Listing 3.9: GLSL tessellation control shader to create an image-space quad with no subdivison

```
1   #version 430
2   layout(vertices = 1) out;
3   flat in vec3 vertColor[];
4   flat in vec3 vertMinPos[];
5   flat in vec2 vertMaxPos[];
6   patch out vec3 controlColor;
7   patch out vec4 controlPosRad;
8   patch out vec3 minPos;
9   patch out vec2 maxPos;
10  void main() {
11    controlPosRad = gl_in[0].gl_Position;
12    controlColor = vertColor[0];
13    if (gl_InvocationID == 0) {
14      gl_TessLevelInner[0] = 1.0;
```

```
15    gl_TessLevelInner[1] = 1.0;
16    gl_TessLevelOuter[0] = 1.0;
17    gl_TessLevelOuter[1] = 1.0;
18    gl_TessLevelOuter[2] = 1.0;
19    gl_TessLevelOuter[3] = 1.0;
20    minPos = vertMinPos[0];
21    maxPos = vertMaxPos[0];
22  }
23  gl_out[gl_InvocationID].gl_Position
24    = vec4(gl_in[0].gl_Position.xyz, 1.0);
25 }
```

Listing 3.10: GLSL tessellation evaluation shader selecting quad corners (cf. Listing 3.5)

```
1  #version 430
2  layout(quads, equal_spacing, ccw) in;
3  patch in vec3 controlColor;
4  patch in vec4 controlPosRad;
5  patch in vec3 minPos;
6  patch in vec2 maxPos;
7  layout(location = 0) flat out vec3 geoColor;
8  layout(location = 1) flat out vec4 geoPosRad;
9  void main() {
10   geoPosRad = controlPosRad;
11   geoColor = controlColor;
12   gl_Position.x = (gl_TessCoord.x < 0.5) ? minPos.x : maxPos.x;
13   gl_Position.y = (gl_TessCoord.y < 0.5) ? minPos.y : maxPos.y;
14   gl_Position.z = minPos.z;
15 }
```

The only difference between the tessellation evaluation shader (cf. Listing 3.10) and the vertex shader segment in the quad-based approach (cf. Listing 3.5) is in use of the variable to select the quads corner. The tessellation evaluation shader uses the tessellation coordinate for this task, which is a parameteric representation of the patch.

CHAPTER 4

Acceleration Strategies

Chapter 3 presented several options for the different calculations required for particle-based ray casting. These include different approaches of glyph silhouette approximation and support geometry generation. Basically, all the different approaches shift workload between different stages of the graphics pipeline, i.e., between geometry upload on the PCIe bus, vertex processing, geometry assembly, rasterization, and fragment processing.

Assuming dynamic data, e.g., from a live coupled simulation, full data transfer is required in every frame. The maximum performance is thus limited by the memory bus speed. In Section 3.3.4 we discussed different strategies to limit the transfer to a single vertex per particle. If data cannot be stored in GPU memory, approaches using more vertices per particle are detrimental. Section 4.1 details different techniques of vertex data upload.

Assuming the data is available on the GPU, the silhouette approximation and actual ray casting calculations remain to be optimized. The different techniques basically trade off load between vertex processing and fragment processing. Additional relevant stages of the GPU include the rasterizer and fragment operations. However, as these stages are not programmable there is not much optimization potential. Instead, we can just make sure to reduce their load as best possible. The question remains which stage is expected to be the primary bottleneck and should be optimized: silhouette approximation in the vertex stage or ray casting in the fragment stage? This question cannot be answered in general. As a rule of thumb, the answer depends on the number of glyphs and the image-space size these glyphs will cover. On the one hand, when rendering tens or hundreds of millions of glyphs, the data transfer will likely be the most severe bottleneck. Between the remaining two stages, the vertex processing will be more critical than the fragment operations. For each particle some size needs to be assumed. Most particles will be the size of a single pixel or even smaller. Thus, there will only be a single ray cast for each glyph. Therefore, in Section 4.2 we detail different approaches for the vertex processing.

On the other hand, when glyphs remain large in image space, the ray casting operations will dominate the calculation times. This is especially true for complex glyphs, like compound glyphs or glyph shapes without analytic ray casting solution. Thus we like to limit their invocation as much as possible, i.e., restrict them to only the fragments contributing to the final image. Section 4.3 presents an approach to exactly achieve this optimization, at the expense of additional data structures.

To summarize the preliminary considerations, there will be no optimal solution. A particle rendering code can either be optimized for one single application. Or we can aim for a compromise performing well in almost all cases, better in some, worse in others. The following subsections present techniques for both approaches, and present a discussion on when to use which technique.

4.1   OPTIMIZED DATA UPLOAD

First we try to find the optimal solution for the particle data upload to the GPU. We assume a streaming scenario, i.e., no data can be stored on the GPU and reused several times. If this is an option, the presented cases will only become easier. The following subsections will hint at corresponding possibilities. We further assume that we want to upload only a single vertex per particle. And we assume, for now, that fragments are cheap to render, in terms of the fragment shader as well as of further fragment operations.

4.1.1   VERTEX ARRAYS

The fundamental approach to vertex data streaming is the usage of vertex arrays. These date back to the time before vertex buffer objects were available and allowed to issue grouped draw commands, remedying issues of the immediate mode rendering and display lists. Listing 4.1 shows an example host code issuing a draw call for multiple GL_POINTs using data transfer via vertex arrays.

Listing 4.1: Host code issuing a draw call for multiple points using vertex arrays for the data transfer

```
 1   // we have color (RGB bytes) and position (XYZ floats) information
 2   // Color is uploaded as generic vertex attribute
 3   GLuint colAryIdx = ::glGetAttribLocationARB(/*shader id*/, "colData");
 4   glEnableVertexAttribArrayARB(colAryIdx);
 5   glVertexAttribPointerARB(colAryIdx, 3, GL_UNSIGNED_BYTE, true,
 6     0 /*stride*/, colorData);
 7   // The position is uploaded with the dedicated function
 8   glEnableClientState(GL_VERTEX_ARRAY);
 9   glVertexPointer(3, GL_FLOAT, 0 /*stride*/, vertexData);
10   // issue drawing for 'cnt' points
11   glDrawArrays(GL_POINTS, 0, cnt);
```

Note that the color is uploaded as generic vertex attribute in line 5. We could use glColorPointer instead as well. One difference is that we use modern shader codes and want to avoid using the built-in shader variables as much as possible. glColorPointer binds to the built-in vertex shader input variable gl_Color. Using a generic vertex attribute we can map the data to any shader variable we like, colData in this case. The same is true for the use of glVertexPointer, which binds to the deprecated shader input variable gl_Vertex. We use this function here to show that the differences in syntax are minimal. One thing missing in most dedicated functions is the boolean *normalize* parameter. Its value controls if data is transferred to the shader as is, or if the data will be rescaled based on the data type. In the case of the color in line 5, the data is uploaded as bytes, i.e., with data in $[0, 255]$. In the shader, this data is accessed as float vector. Using normalization, the fourth parameter, the values are scaled from $[0, 255]$ to $[0, 1]$.

Table 4.1: Different data layouts and the resulting stride values. Vertex position data is assumed to be floats. Color data is assumed to be bytes.

data stream	vertex stride	color stride
XYZXYZXYZ...	12 (or 0)	
RGBRGBRGB...		3 (or 0)
XYZRGBXYZRGB...	15	15

Another interesting parameter is *stride*, which was set to zero in the above example. Stride is the number of bytes from the start of one data element to the next one. Speaking of the position information of 3D vectors of floats and assuming the data is tightly packed as continuous array, this distance would be 12 bytes. As a special value, a stride of zero, used in the example above, specifies a tightly packed array, i.e., the element size (e.g., 12 bytes) is automatically computed based on the data type (e.g., parameters 2 and 3 in line 5, or 1 and 2 in line 9 of Listing 4.1). Table 4.1.1 shows three examples of data stream layouts and the corresponding stride values. Example 3 shows the possibility to define an interleaved data layout of different vertex attributes. Listing 4.2 shows the corresponding function calls.

Listing 4.2: Function call defining interleaved vertex attributes similar to Listing 4.1

```
1  // per-byte addressing irrespective of actual data types
2  // uint8_t* allData;
3  glVertexAttribPointerARB(colAryIdx, 3, GL_UNSIGNED_BYTE, true,
4    15 /*stride*/, allData + 12);
5  glVertexPointer(3, GL_FLOAT,
6    15 /*stride*/, allData);
```

Note that the data point needs to be of bytes-type for the pointer arithmetic in line 3 to work correctly. The performance measurement in the previous publication by Grottel et al. [2009a] showed that interleaved data layout is beneficial on some graphics cards and in some scenarios. On modern graphics cards the effect is usually not present anymore. The ability to easily define different data layouts shows the expressiveness of this simple API.

The functions to specify the data points can also be used to bind vertex buffer objects to vertex attributes, cf. Section 4.1.2. In these cases the pointers (last parameters) specify the byte offset relative to the beginning of the vertex buffer object, i.e., usually zero. The use of these functions for vertex arrays, i.e., without activated vertex buffer object, however, has been marked deprecated in modern (forward-compatible) OpenGL. The recommended upload strategy in modern application uses vertex buffer objects with accordingly set attributes.

4.1.2 VERTEX BUFFER OBJECTS

Vertex buffer objects represent GPU memory which will be used for vertex attributes. They are allocated on the GPU, and data is uploaded in a generic way. Then the buffer objects are bound to vertex attributes. Here the interpretation of the data stored is specified, i.e., the data type and number of components per element. In modern OpenGL, these bindings to vertex attributes and the data specification is stored in a vertex array object. Obviously, this approach limits the data to the amount of free GPU memory. However, data can always be uploaded and rendered in chunks.

In accordance with the streaming scenario, our rendering code requires three tasks to be fulfilled: initialization, data upload, and rendering. The initialization creates the required buffer objects and the bindings to the vertex attributes. Listing 4.3 shows a corresponding host code.

Listing 4.3: Initialization of vertex buffer objects for particle rendering

```
1  glGenBuffers(1, &theVBO);
2  glGenVertexArrays(1, &theVAO);
3
4  glBindVertexArray(theVAO);
5  glEnableVertexAttribArray(0); /* position */
6  glBindVertexBuffer(0, theVBO, 0, 4 * 3);
7  glVertexAttribFormat(0, 3, GL_FLOAT, GL_FALSE, 0);
8  glVertexAttribBinding(0, 0);
9  glBindVertexArray(0);
```

The fourth parameter in line 6 is again the stride value, explicitly specified this time. `glVertexAttribFormat` specifies the data format similar to `glVertexAttribPointerARB` in Listing 4.1 but without the last parameter (the data pointer or data offset). The colors are used to differentiate the meanings of all the zeros:

In orange the vertex array object is created in line 2, activated for setup in line 4, and deactivated again in line 9. The explicit deactivation is not required in modern OpenGL, as states are seldom deactivated, but just changed. However, explicitly deactivating states (or state objects) is basically defensive programming to make sure settings within these objects are not accidentally changed. The blue text marks the vertex buffer object actually storing the data.

In green the location of the vertex attribute variable is marked. We explicitly use location zero and assume the corresponding shader uses this location for the matching input variable, e.g., specifying it with the corresponding `layout(location = 0)` modifier. Line 5 enables the array data access for this variable. Otherwise the corresponding value from the OpenGL state would be used (cf. `glVertexAttrib`). Line 7 specifies the data layout, matching the variable definition in the vertex shader code.

In red a vertex buffer binding point is specified. These binding points represent active or selected vertex buffer objects to be used when drawing in this setup. Line 6 binds the vertex buffer object to one such binding point, i.e., zero, and specifies data offset and element stride. The final connection between the vertex buffer data and the vertex attribute is specified in line 8. This, similar to the vertex array functions, allows us to use different vertex buffer objects for different vertex attributes, or to specify different (subsets of) vertex attributes stored interleaved in the same vertex buffer object, using the stride and offset values. This concludes the setup.

To upload data to the vertex buffer object, we need to be aware that there was no memory allocation yet. Data can be uploaded replacing any existing data. This usually requires the GPU memory management to release the data buffer and allocate a new one, costly operations which can be avoided if it is known that the required buffer size did not change. This concept has been known for textures for a long time and is directly transferable to vertex buffer objects as well. Listing 4.4 shows the corresponding host code.

Listing 4.4: Data upload into a vertex buffer object using *direct state access* (aka bind-less)

```
1  glBindVertexArray(theVAO);
2  if (firstFrame) {
3    glNamedBufferDataEXT(theVBO,
4      4 * 3 * cnt, data,
5      GL_STATIC_DRAW);
6  } else {
7    glNamedBufferSubDataEXT(theVBO, 0,
8      4 * 3 * cnt, data);
9  }
```

The functions in this example use direct state access, i.e., directly specify the vertex buffer object by its name, not requiring it to be actively bound to anything and especially not changing the OpenGL state. The first function version in line 3 is used for initial data specification and upload. The second version in line 7 is used to replace the stored data. While this function allows us to replace parts of the data, we replace the whole data, still reusing the allocated buffer object. Note that the usage hint `GL_STATIC_DRAW` is only specified when allocating the object.

To finally draw the data only the vertex array object must be activated. All correspondingly set up vertex buffer object and vertex attribute connections are retrieved. The actual draw code is thus reduced to the lines shown in Listing 4.5.

Listing 4.5: Drawing point data from vertex buffer objects specified with a vertex array object.

```
1  glBindVertexArray(vertArrays);
2  glDrawArrays(GL_POINTS, 0, cnt);
3  glBindVertexArray(0);
```

The diagrams in Figure 4.1 compare the performance of the rendering approaches using vertex arrays and vertex buffer objects. The left diagram shows the overall rendering timings including upload and simple splatting for constant colored, single-fragment sized primitives. *VA* marks the data for rendering with vertex arrays. *VBO* marks the data for rendering with vertex buffer objects. The additional specifier for this implementation details the *usage* setting defined when creating the vertex buffer object, either `GL_STATIC_DRAW`, `GL_STREAM_DRAW`, or `GL_DYNAMIC_DRAW`. *Su-*

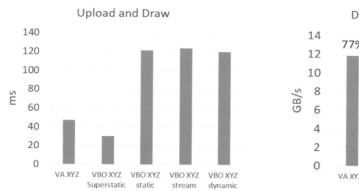
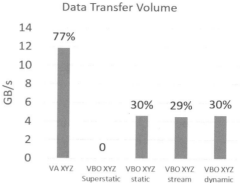

Figure 4.1: Performance diagrams comparing rendering with vertex arrays against rendering with vertex buffer objects; left: rendering times containing data upload; right: data upload performance and PCIe utilization percentage.

perstatic marks a measurement where the data is uploaded into a vertex buffer object with the usage setting `GL_STATIC_DRAW` and is never updated again. Thus, this is really static data and shows the performance base line of the fixed rendering overhead. The data set used contained 50 million points represented using $3\times$ floats for positions. The system for the measurements was equipped with an Nvidia GeForce GTX 680 GPU and an i7-3770K@3.5GHz CPU. Rendering resolution was set to 1024^2 pixels. The right diagram in Figure 4.1 shows the upload performance computed from the amount of data and the rendering speeds. The percentage written atop of each bar is the utilization of the theoretical maximum capacity of the PICe 3.0 $\times$16 bus.

These results clearly show that vertex arrays are superior over vertex buffer objects. More so, the difference in performance in the vertex buffer objects' usage settings are not significant. This result differs from previous measurements performed by Grottel et al. [2009a]. We believe that this usage scenario does not provide enough possibilities for the GPU (or driver) to optimize data access or locking. As for the difference between vertex arrays and vertex buffer objects, we believe that one of the reasons for the better performance of vertex arrays is the interleaved nature of data upload and rendering. With vertex buffer objects, the data is uploaded to the GPU as a whole, and only after this upload has finished, the actual rendering can take place. In contrast to that, using vertex arrays the upload starts with the draw call. Thus the rendering runs in parallel and can start as soon as the data for the very first vertex is available to the GPU. Seeing these figures makes the choice of marking rendering with vertex arrays as deprecated in the core OpenGL specification somewhat questionable. However, there are further options for data upload to be explored, namely shader storage buffer objects and double-buffered upload.

4.1.3   SHADER STORAGE BUFFER OBJECTS

Shader storage buffer objects are a generic GPU memory object, which are arbitrarily accessible from any shader stage. They are extremely flexible and can be adjusted in many ways to specific purposes. Especially interesting is the possibility to explicitly specify synchronization methods. This way, unnecessary synchronizations, which would only slow down the rendering, can be disabled.

As with vertex buffer objects, shader storage buffer objects can receive their data via different approaches. One is to use memory mapping, i.e., mapping the corresponding GPU memory into the host memory address space. All memory access to this address block will transparently move data to or from the GPU via DMA. Preliminary tests have shown that this approach is superior over the alternative upload commands, especially when the memory is mapped persistently, i.e., is never unmapped as long as the application runs. However, mapped memory requires the underlying pages to be non-swappable (*pinned* or *locked*) and thus affects the system performance if not used in moderation and in sensible relation to the total available memory. More precisely, we cannot allocate very large memory blocks this way. We therefore introduce the concept of interleaved upload and rendering using a double-buffered shader storage buffer object. We use one half of the buffer object as target for the upload, and use the second half as source for a rendering command. Then we synchronize both operations and swap the meaning of the two half buffers. Listing 4.6 shows the host code of the initialization of the buffer objects.

Listing 4.6: Initialization of shader storage buffer objects

```
1  glGenVertexArrays(1, &vertArray);
2  glBindVertexArray(vertArray);
3  glGenBuffers(1, &theBuffer);
4  glBindBuffer(GL_SHADER_STORAGE_BUFFER, theBuffers);
5  glBufferStorage(GL_SHADER_STORAGE_BUFFER,
6    bufferSize * numBuffers, nullptr, bufferCreationBits);
7  void *mappedMemory = glMapNamedBufferRangeEXT(theBuffers, 0,
8    bufferSize * numBuffers, bufferMappingBits);
9  glBindBuffer(GL_SHADER_STORAGE_BUFFER, 0);
10 glBindVertexArray(0);
```

The vertex array object `vertArray` is needed to manage the actual buffer object, although we will use no vertex buffer objects. The size of the buffer itself is specified by `bufferSize * numBuffers`. `numBuffers` is the number of buffers we want to use to interleave upload and rendering. For double-buffering this is 2, but we experimented with more buffers (cf. results below). `bufferSize` is the size of a single buffer block in bytes. We experimented with different sizes here as well.

More interesting are the two variables, `bufferCreationBits` and `bufferMappingBit`, which specify settings for the buffers, most importantly the synchronization set up. We use the `GL_MAP_PERSISTENT_BIT` to inform OpenGL that the buffer will be mapped persistently. GL_MAP_WRITE_BIT specifies that we are writing data into the buffer, i.e., onto the GPU. Note, we do not set GL_MAP_READ_BIT, meaning we will never read data back from the GPU, allowing the GPU cache to be optimized accordingly. The third and final set bit is GL_MAP_FLUSH_EXPLICIT_BIT. With this value we define that we will explicitly request a synchronization when the data transfer must have finished. We will request this synchronization before we swap the meaning of the two (or more) buffers. The mapped memory is accessible via the pointer in line 7.

Listing 4.7: Drawing particles with shader storage buffer objects persistently mapped into memory

```
1   glBindVertexArray(vertArray);
2
3   vertStride = 3 * sizeof(float);
4   numVerts = bufferSize / vertStride;
5   currVertX = static_cast<const char *>(data);
6   vertCounter = 0;
7   while (vertCounter < count) {
8     void *mem = static_cast<char*>(mappedMemory)
9           + bufferSize * currentBuffer;
10    const char *whence = currVertX;
11    GLsizei vertsThisTime = std::min(count - vertCounter, numVerts);
12
13    // wait until this buffer is rendered!
14    waitSignal(fences[currentBuffer]);
15
16    memcpy(mem, whence, vertsThisTime * vertStride);
17    glFlushMappedNamedBufferRangeEXT(theBuffer,
18          bufferSize * currentBuffer, vertsThisTime * vertStride);
19
20    glBindBufferRange(GL_SHADER_STORAGE_BUFFER,
21      0, theBuffer, bufferSize * currentBuffer, bufferSize);
22    glDrawArrays(GL_POINTS, 0, vertsThisTime);
23
24    queueSignal(fences[currentBuffer]);
25
26    currentBuffer = (currentBuffer + 1) % numBuffers;
27    vertCounter += vertsThisTime;
28    currVertX += vertsThisTime * vertStride;
29  }
```

Line 4 computes the number of particles which can be stored in one buffer. vertCounter stores the number of particles which already have been rendered. currVertX is the pointer to the CPU-side memory storing the next particles to be uploaded and rendered. The while loop starting at line 7 continues to upload and render data in blocks until all the data has been rendered. In line 8 mem is the pointer to the mapped GPU-buffer to be uploaded next. In line 11 the number of particles uploaded in this iteration is stored.

First in line 14 the loop waits for the current buffer to be available for writing. In the very first iteration, the fence is open and the thread directly continues with its execution. In line 16 the actual memory transfer is initiated. Line 17 forces the write (to GPU) to be completed, being the second synchronization point. Thus data transfer is now completed.

In line 20 the current buffer range the new data has been written to is mapped to the buffer binding point 0 and will now be available for the GLSL shaders. The subsequent draw call initiates the rendering of the uploaded particles. Note that OpenGL rendering is inherently asynchronous. glDrawArrays returns very quickly and only a following and conflicting OpenGL command might block until the drawing is actually completed. For our approach, we need to know when the rendering is completed to synchronize (cf. line 14). In line 24 we call a utility function to issue a synchronization point into the OpenGL command stream. Details on the synchronization functions is given by their implementation in Listing 4.8. In lines 26–28 all iterating pointers and counters are incremented.

Listing 4.8: Utility functions to queue a synchronization fence into the OpenGL command stream, and to wait for this fence. waitSignal will return only after the command stream completed all previous commands including the fence.

```
1   void queueSignal(GLsync& syncObj) {
2     if (syncObj) glDeleteSync(syncObj);
3     syncObj = glFenceSync(GL_SYNC_GPU_COMMANDS_COMPLETE, 0);
4   }
5
6   void waitSignal(GLsync& syncObj) {
7     if (!syncObj) return;
8     while (1) {
9       GLenum wait = glClientWaitSync(syncObj,
10        GL_SYNC_FLUSH_COMMANDS_BIT, 1000);
11      if (wait == GL_ALREADY_SIGNALED || wait == GL_CONDITION_SATISFIED)
12        return;
13    }
14  }
```

The utility function queueSignal simply inserts a synchronization point into the OpenGL command stream, which will be activated as soon as the OpenGL command processing reaches this point. More interestingly, waitSignal actively waits for the synchronization point to be set. The surrounding while loop starting at line 8 is needed since glClientWaitSync might return with GL_TIMEOUT_EXPIRED or GL_WAIT_FAILED. The third parameter in glClientWaitSync is the time out value in nanoseconds. The actual value is not very important, since the wait time is complemented by the while loop.

The whole rendering process is sketched in Figure 4.2. In this example we assume that the data is uploaded faster than it can be rendered, i.e., green bars are longer than blue ones. The data is uploaded in four blocks indicated by the numbers in the corresponding upload and rendering

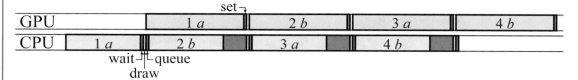

Figure 4.2: Sketch of the interleaved rendering process. Blue blocks show data upload and green blocks show rendering of the corresponding data blocks. The green thin stripes in the CPU band are the rendering calls `glDrawArrays`. Pink stripes represent the synchronization points: waiting on a signal, add a signal to the GPU command stream, and set the signal when reached on the GPU. The two used buffers are indicated by *a* and *b*. In this example rendering is slower than upload.

blocks. The two buffers in the double-buffer approach are indicated by *a* and *b*. Synchronization points are shown in pink. The *wait* points corresponds to line 14 in Listing 4.7, *queue* to line 24, and *set* marks the actual point in time when the command stream on the GPU reaches the fence object.

Listing 4.9: GLSL vertex shader segment fetching position from shader storage buffer object

```
1  layout(packed, binding = 0) buffer shader_data {
2    float theBuffer[];
3  };
4  #define POSITION vec3(theBuffer[gl_VertexID * 3], \\
5    theBuffer[gl_VertexID * 3 + 1], theBuffer[gl_VertexID * 3 + 2])
```

Listing 4.9 shows the beginning of a vertex shader and shows the lines accessing the shader storage buffer object. The `binding = 0` in line 1 is the buffer binding point and corresponds to the 0 in line 21 in Listing 4.7. The whole buffer is interpreted as array of float. Using `gl_VertexID` in lines 5 and 6, the vertex shader fetches the three position float values for its invocation. Note that you can define more complex data structures as layout of the shader storage buffer object. However, you must be very careful considering size and alignment, e.g., each variable needs to be aligned to a multiple of four bytes, making a CPU-side tightly packed combination of float 3D position and byte RGB color invalid. To avoid these problems, you could strictly separate all vertex attributes into different shader storage buffer objects, following the structure-of-arrays approach. This might have a performance impact depending on different graphics cards for better or worse, but has not been tested thoroughly yet.

A performance comparison of all three rendering approaches is shown in the diagrams in Figure 4.3. Again, only constant-color single-fragment points are rendered to restrict the performance variability to the data upload. The interleaved rendering using shader storage buffer objects performs well compared to the use of vertex buffer objects. However, none of these modern approaches is able to reach the throughput available when using vertex arrays. The shader storage

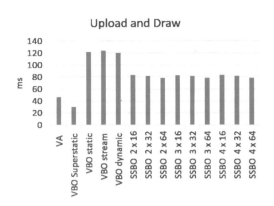

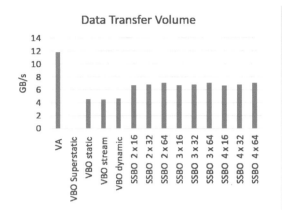

Figure 4.3: Performance diagrams comparing between rendering with vertex arrays, with vertex buffer objects, and with shader storage buffer objects; left: rendering speed containing data upload; right: data upload performance.

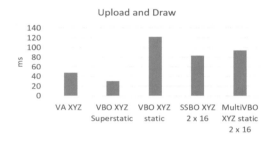

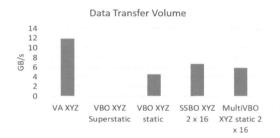

Figure 4.4: Performance diagrams comparing all rendering approaches; left: rendering speed containing data upload; right: data upload performance.

buffer objects were measured with different settings. For one, the number of buffers used was changed from 2 to 3 and 4. Additionally, the buffer size was increased from 16 KB to 32 KB and 64 KB. The performance numbers of these different setups are very similar, and reach 45–50% of PCIe3.0 ×16 bus utilization. In particular, the number of buffers is not relevant at all. Thus the upload is the bottleneck, not the rendering. This might be different when using more demanding shaders or graphics cards with fewer GPU cores but the same memory bus. There always is a slight performance increase when the buffer size is increased. This stands to reason because of the fewer synchronization points and larger DMA block transfers. However, the increase is not significant enough, given the fact the pinned memory should be used sparingly.

One difference between the implemented rendering code using shader storage buffer objects and vertex buffer objects is the double-buffer approach to interleave rendering and upload.

Of course, the question arises if this is the reason for the performance difference. Figure 4.4 summarizes the rendering speed and data transfer bus utilization for all rendering approaches. The very right bars *MultiVBO* show the figures for a rendering implementation using vertex buffer objects in a double-buffer setup. The shader storage buffer objects still perform better, because of the more flexible synchronization mechanisms. The vertex buffer objects are synchronized automatically by OpenGL and the GPU.

With these findings, it is not easy to specify the one best approach. Instead the following rules of thumb should help a graphics developer choose an approach matching his or her needs. *Vertex array*s deliver absolute peak performance in the data streaming. But they are deprecated in modern OpenGL. *Shader storage buffer objects* perform very well, but the synchronization needs to be handled manually to use them optimally. This requires some cumbersome implementation in host code as well as in shader code (cf. data alignment). *Vertex buffer object*s are the easiest to implement because of the fully automatic synchronization. Using them in a simple double buffer setup achieves good perfomance as well. Again, note that these figures only hold if the data upload is the bottleneck.

4.2   SUPPORT GEOMETRY GENERATION

The previous section optimized uploading of points for single-fragment splatting. Fragment processing load was completely negligible. However, for glyph-based visualization, this is not true. Not only do we perform a rather costly ray casting in the fragment shader, we also need to generate multiple fragments for each vertex. Section 3.2 detailed approaches using `gl_PointSize`, Quad upload, the geometry shader and the tesselation shader. Using quads is not really an option when aiming for visualization of large dynamic data. Figure 4.5 shows the performance figures for the geometry generation approaches. For this test 50 million spheres were fully rendered, performing ray casting including local lighting computation. Early z test was disabled. All spheres were visible and had a screen-space size of approximately two to three pixels. The overall rendering viewport size was 1920×1080. The test system was equipped with an Nvidia GeForce GTX 680 GPU and an Intel i7-3770K@3.5GHz CPU.

`GL_POINTS` massively outperform the other techniques. We believe this is because the geometry generation task is still hard wired in the graphics pipeline, and the clipping and rasterization is utterly trivial. Using the *geometry shader* to span a triangle strip with two triangles performs second best, although a rendering time of 121 ms can hardly be called interactive. The worse performance is due to several factors: for one, the vertices are emitted in serial manner in the geometry shader. As already discussed in Section 3.3.4, this however is not too critical, as the geometry shader does not really need to carry out any calculations. We believe the more severe issue is that a further stage in the graphics pipeline is activated, introducing additional synchronization points, caching and buffering tasks. Interestingly, the *tessellation shader* approach performed even worse, although the serial vertex generation is now fully parallelized. This supports our statement that the serial processing in the geometry shader is not the pressing problem, but the additional pipeline stages are. The tessellation shader introduces two new stages (three if you count the fixed-function tessellator itself), while the geometry shader only introduces one.

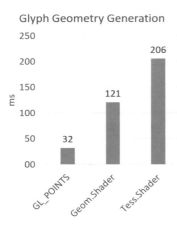

Modern GPUs, more precisely functionality like shader stroage buffer objects, enable one further approach to generate support geometry: attribute-less rendering. Usually, vertex attributes are specified, for example pointing to vertex buffer objects, for at least position data. Then the graphical primitives are defined with type and count to invoke the vertex shader stage the required number of times, and the vertex attributes are automatically fetched and passed to the vertex shader. The vertex attributes, however, are purely optional. While this is understandable for all additional attributes, like color, normals, texture coordinates, and the like, it is not obvious for the positional attribute. Vertex shaders are only required to write gl_Position, otherwise the pipeline stops. There are no requirements on the input variables. Using shader storage buffer objects, a vertex shader can fetch the required data itself. Which data to fetch can be selected, e.g., using gl_VertexID. The idea here is to generate quads (two triangles) for single particles, to upload the minimal particle data into shader storage buffer objects, and then to issue a draw command for two triangles per particle, i.e., for six vertices per particle: glDrawArrays(GL_TRIANGLES, 0, count * 6);.

Figure 4.5: Performance of difference geometry generation approaches.

Listing 4.10: Vertex shader excerpts fetching particle data and selecting image-space corner based on the gl_VertexID

```
1  layout(packed, binding = 0) buffer shader_data {
2    float theBuffer[];
3  };
4  #define POSITION vec3(theBuffer[(gl_VertexID / 6) * 3], \\
5    theBuffer[(gl_VertexID / 6) * 3 + 1], \\
6    theBuffer[(gl_VertexID / 6) * 3 + 2])
7
8  // ... vertex shader, e.g., similar to listing 3.4
9
10 gl_Position.x = ((gl_VertexID % 2) == 0) ? minPos.x : maxPos.x;
11 int rmdr = gl_VertexID % 6;
12 gl_Position.y = (rmdr > 3 || rmdr == 2) ? maxPos.y : minPos.y;
13 gl_Position.z = minPos.z;
14 gl_Position.w = 1.0;
```

Listing 4.10 shows two segments of a vertex shader. Lines 1–6 fetch the positional information of the respective particle, i.e., six vertices for the two triangles select all the same vertex.

Loading further data, e.g., color, would be addressed accordingly. In the omitted center part of the shader, the image-space silhouette is estimated and stored in the variables `minPos` and `maxPos`. Lines 10–12 select the correct corner from these positions, based on the vertex id.

Given performance numbers in Figure 4.5, the rendering approach with `GL_POINTS` and `gl_PointSize` is obviously favorable. The only disadvantage is the vendor specific point-size limit. If this is an issue, the geometry shader provides a viable solution. An optimized rendering system might even use both techniques. Based on a spatial data structure, e.g., a simple grid, the maximum glyph image-space size can be estimated for whole groups of particles. If this size exceeds the maximum size for points for some groups, these particles can be rendered using the geometry shader. Usually, this will only be the case for relatively few particles close to the camera. All remaining points, i.e., most of the data set, will be small enough for the point-based rendering approach. The attribute-less rendering reaches roughly the same performance as when using the geometry shader. Given the fact that the vertex computation load is increased by a factor of six, this approach remains interesting, but currently not really usable. Of course, the support geometry could be reduced to a single triangle, reducing the vertex computation load, however, then many more unnecessary fragments would be created, thus potentially shifting the bottleneck to the fragment stage. The tessellation shader approach is very slow. So the two most modern technologies are currently not viable options. However, the same was true a few years ago when talking about the geometry shader. Therefore, these approaches remain interesting for future GPUs.

4.3   PARTICLE CULLING TECHNIQUES

Given the potential of glyph ray casting and data transfer is depleted with the methods already described, to interactively visualize ever bigger data sets the only option left lies in optimizing the rendering process as a whole. Comparing the data set sizes with the limited resolution and size of computer displays, obviously the particle/pixel ratio suggests that there will be many unnecessary fragments which will be overwritten. Considering a standard computer display has roughly 2 million pixels (e.g., Full HD: 1920×1080 pixels), only that many particles can be visible at most. Transferring and ray casting that many particles does not pose a challenge for current hardware. So the optimization strategy is to remove as many particles as possible from the active data which will not result in visible pixels anyway. This is the idea of the computer graphics concept known as (frustum or occlusion) culling. The according method was originally presented in Grottel et al. [2010a].

4.3.1   OCCLUSION QUERIES FOR GRIDDED DATA

The main idea is to use a two-stage culling to reduce both the data transfer as well as the rendering load. Figure 4.6 summarizes the different stages of this culling technique. The data transfer load is reduced by determining visibility of spatial subvolumes of the data using hardware-supported occlusion queries. Hardware occlusion queries determine the number of visible fragments created by rasterizing a geometry against the depth buffer. The rendering is further optimized using

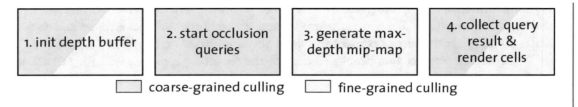

Figure 4.6: Schematic process steps of a two-stage culling method for particle-based data. Generating the depth mipmap in stage 3 is required for fine-grain culling during stage 4 and hides the latencies introduced by the occlusion queries issued in stage 2.

a replacement for the unavailable early depth test using a hierarchical maximum-depth mipmap. Details on the manual early depth test will be described in Section 4.3.2. Using hardware occlusion queries introduces some latency, due to the asynchronous nature of the GPU. The latency introduced by the queries is hidden by performing operations for the maximum-depth mipmap in parallel.

The presented two-level occlusion culling builds on the following stages which are also outlined in Figure 4.7:

1. Initialize the depth buffer for occlusion culling rendering cheap conservative depth estimates for the particles which were visible in the last frame.
2. Issue hardware occlusion queries for bounding boxes of all cells of the spatial data structure.
3. Compute a maximum-depth mipmap from the depth buffer of step 1 for fine-granular culling.
4. Perform final rendering.
    4.1. Read back the results of the hardware occlusion queries, update the list of visible particles, and render all visible glyphs with per-glyph culling directly on the GPU.
    4.2. Perform deferred shading. Details are given in Section 7.1.

The hardware occlusion queries require a coarse subdivison of the data. We use a regular grid as spatial structure to organize the particle data. A hierarchical data structure is not reasonable, because it would make a stop-and-wait algorithm necessary, as one hierarchy level would need the results of the previous one. Solving this issue in an efficient manner is complex and not needed.

Step 1 of the presented method renders all particles from the grid cells marked as previously visible into the depth buffer for the current frame. The resulting depth buffer will be used to perform both types of culling. For these glyphs, the ray casting algorithm is not performed in its complete form. Instead only a conservative maximum depth estimate is calculated in the vertex shader and is not changed in the fragment shaders, thus enabling the hardware-supported early depth test. The efficiency of the depth test is further enhanced by sorting the cells to be rendered from front to back, which can be carried out very quickly by using a stable sorting algorithm and

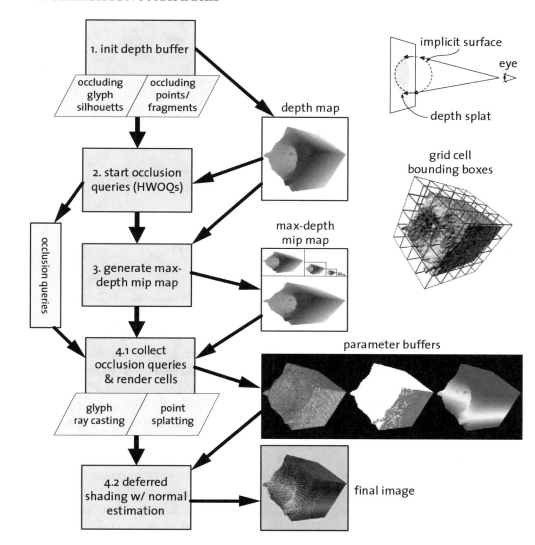

Figure 4.7: Details on the stages of the culling method: 1. Initialization of the depth buffer with known occluders from the previously rendered frame; 2. Start of hardware occlusion queries for all grid cells by testing their bounding boxes; 3. Generation of maximum-depth mipmap; 4.1. Collection of results of hardware occlusion queries, update of the list of visible cells, and rendering of visible glyphs. Stages 1 and 4.1 can output ray cast glyphs or single flat-shaded fragments if the glyphs become too small in image-space. Stage 4.2 implements deferred shading and is described in detail in Section 7.1. Note that the rendering in stage 1 initializes the depth buffer with a conservative depth splat for the maximum-depth mipmap, as well as for subsequent render passes.

keeping the order across frames. The fragment shader performs only the calculation to generate the precise silhouette of the glyph.

Listing 4.11: Issuing occlusion queries

```
1   glColorMask(GL_FALSE, GL_FALSE, GL_FALSE, GL_FALSE);
2   glDepthMask(GL_FALSE);
3   glDisable(GL_CULL_FACE);
4   // also disable texturing and any fancy shading features
5   for (int i = cellcnt - 1; i >= 0; --i) { // front to back
6     CellInfo& info = cellinfos[i];
7     if (!info.boundingBox.isvisible) continue; // frustum culling
8     glBeginQuery(GL_SAMPLES_PASSED, info.oQuery);
9     // render bounding box for cell idx
10    glEndQuery(GL_SAMPLES_PASSED);
11  }
12  glColorMask(GL_TRUE, GL_TRUE, GL_TRUE, GL_TRUE);
13  glDepthMask(GL_TRUE);
14  glEnable(GL_CULL_FACE);
15  // reenable other state
```

In stage 2 the hardware occlusion queries are issued against the created depth buffer by rendering the bounding box geometries for all cells of the grid. Listing 4.11 shows code excerpts issuing occlusion queries. Note that `info.oQuery` is a query object created with `glGenQueries`. Cells previously invisible might now become visible and vice versa because of view-dependent changes in particle densities within the occluding cells or changes of the view parameters. The performance of this stage depends on the amount of issued hardware occlusion queries which result from the resolution of the spatial grid. On the one hand, more cells will yield more accurate results for the cell-level culling, but on the other hand, will also create more queries and thus longer latencies until the query results are available for stage 4.1. For example, using a grid of 15^3 cells requires 3,375 occlusion queries. The best grid resolution depends on the graphics hardware as well as the data set structure and, thus, needs to be adjusted for each scenario anew.

The task of coarse-grain culling on cell-level is completed in stage 4 in which the results of the hardware occlusion queries are read back from the graphics hardware. This value is returned by the command `glGetQueryObjectui64v(info.oQuery, GL_QUERY_RESULT, &fragCnt);`. The visibility flags of the cells of the spatial grid are then updated correspondingly and the particles of the visible cells are finally rendered using ray casting. Using a conservative estimate based on the particle size and cell placement relative to the viewpoint, the rendering code can estimate the maximum image-space size of the particles within a cell. If particles are too small, i.e., of the same size as single pixels, ray casting is not required and simple splatting results in acceptable image quality. See Section 7.1 for details on a corresponding image-space pass. This concludes the coarse-grain culling reducing the data transfer.

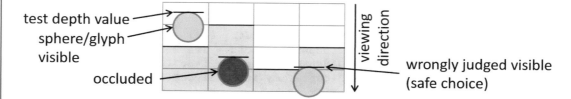

Figure 4.8: Replacement of unavailable early depth test. Glyphs depths value is estimated in the vertex shader (indicated by lack lines above spheres). The maximum-depth mipmap texture is queried at corresponding mipmap levels. Thus only four texel fetches are required.

4.3.2 MANUAL EARLY-Z TEST

Current GPUs in principle already support early depth tests with hierarchical maximum-depth mipmaps natively. However, glyph-based ray casting disables this mechanism by having the fragment shader overwrite the depth value estimated by the rasterization stage. Obviously, as soon as a fragment shader writes a depth value, the GPU scheduler must assume any depth value is possible, and therefore disables the early depth test. There is a relatively recent extension to this function, the conservative depth. In the fragment shader a layout directive informs the graphics pipeline that the depth will only be changed in certain ways. We still explain our implementation to illustrate the concept in detail.

We thus implemented a manual fine-grain culling as replacement. As mentioned in the previous section, for this the maximum-depth mipmap needs to be computed. Stage 3 performs this calculation based on the depth buffer created in stage 1. A mipmap pyramid can be created via Ping-Pong rendering with two frame buffer objects (FBOs), down-sampling using the maximum of 2×2 values for the next mipmap level. Note that this operation can also be performed entirely on the GPU using a compute shader. Section 7.2 holds a corresponding example used in another context. However, if a graphics card is used with compute shader capabilities, it will also support conservative depth and thus this early depth test replacment will not be needed anyway.

Listing 4.12: GLSL vertex shader segment evaluating maximum-depth mipmaps to manually perform an early depth test

```
1   // depth texture coordinates ..
2   vec2 dtc = gl_Position.xy + vec2(1.0);
3   dtc /= vec2(viewAttr.z, viewAttr.w);
4
5   int miplevel = min(max((int(log2(gl_PointSize))), 1), DEPTHMIP_MAXLEVEL);
6   float exp = exp2(float(miplevel));
7   dtc /= exp;
8   ivec2 idtc = ivec2(dtc - vec2(0.5)); // cast to ivec2 performs a "round"
9   // now in relative coordinate of the mip level
10  idtc.x += int(float(DEPTHMIP_WIDTH * (1.0 - 2.0 / exp)));
```

```
11  idtc.y += DEPTHMIP_HEIGHT;
12
13  vec4 depth1 = texelFetch2D(depthTex, idtc, 0);
14  vec4 depth2 = texelFetch2D(depthTex, idtc + ivec2(1, 0), 0);
15  vec4 depth3 = texelFetch2D(depthTex, idtc + ivec2(0, 1), 0);
16  vec4 depth4 = texelFetch2D(depthTex, idtc + ivec2(1, 1), 0);
17  float depth = max(max(depth1.x, depth2.x), max(depth3.x, depth4.x));
18  if (depthPos.z > depth) {
19    gl_Position.w = 0.0;
20  }
```

When rendering particles in stage 4.1 the image-space size of the glyph is estimated. This estimation also yields the mipmap level we need to query whether the glyph is occluded or not. The concept is illustrated in Figure 4.8. In the correct level, the glyph's image-space footprint covers four texels at most. Listing 4.12 shows the corresponding segment of a vertex shader. Note that size and maximum level of the depth mipmap texture is specified with defines (written in upper case). These values can be uploaded using a uniform variable. The maximum value of these is compared to the minimum depth of the glyph to decide the glyph's visibility in a conservative way. Glyphs, which are definitely hidden, are removed in their vertex shader by being moved to infinity or in front of the near clipping plane. While the geometry processing and vertex shaders are still executed for all glyphs transferred to the graphics card, the costly rasterization and ray casting in the fragment shader is only performed for the particles which are potentially visible.

The performance of the manual early-z test depends on the location of the bottleneck. If the actual rendering speed is limited by the fragment load, this test can increase the rendering performance by an order of magnitude. For large data the combination of the coarse-grain culling (cf. Section 4.3.1) and this fine grain culling is especially beneficial. Detailed results can be found in the original publication by Grottel et al. [2010a].

CHAPTER 5

Data Structures

The acceleration strategies described in Chapter 4 enable us to render several hundreds of thousands of particles at interactive rates. However, these techniques are only profitable for dense data featuring inherent occlusion and, thus, potential for culling. When dealing with even more particles or sparse data, we need to resort to explicit data structures for both storing the data on the GPU as well as efficient data access during rendering.

The first part of this chapter (cf. Section 5.1) will deal with block-based data structures, in particular uniform grids. This type of data structure fits nicely with the current GPU architectures and is straightforward to implement and use through texture mapping functionality. In the context of this book, uniform grids are used to speed up the rendering of billions of atoms at the cellular scale and, at the same time, to save memory by instancing. The key to this technique is the fact that the cellular environment, although containing billions of atoms, consists of lots of similar structures [Falk et al., 2013b, Lindow et al., 2012]. This section also details the integer-based traversal of uniform grids presented by Amanatides and Woo [1987] adapted to GLSL.

In Section 5.2, we discuss hierarchical data structures for a more general semantic abstraction [Le Muzic et al., 2014]. The main application for this approach is the visualization of proteins with a dynamic level of detail. We conclude this chapter with a brief discussion on the work by Hopf and Ertl [2003] on position quantization. Its underlying data structure is a hierarchy obtained by clustering the particle data. The quantization of the coordinates can reduce the memory footprint of particle data while still maintaining a high level of accuracy.

5.1 UNIFORM GRIDS FOR MOLECULAR DYNAMICS DATA

Uniform grids are a basic data structure for the spatial subdivision of an n-dimensional Euclidean space. The size of the individual grid cells is constant along each dimension, hence the name uniform grid. The main advantage of this data structure is that one-, two-, and three-dimensional grids can directly be mapped onto 1D, 2D, and 3D textures on the GPU. In addition, the GPU provides sampling and basic filtering of the data. The traversal of the grid also does not require any special means or computationally expensive calculations. On the other hand, the grid might contain a varying number of particles per cell depending on the spatial distribution of the data and the data structure does not encode any hierarchical information.

Since existing texture formats support only up to four data values per texel texture indirection has to be used to be able to store and access more data values. Here, the texels or voxels of the source texture, which represents the uniform grid, point to additional textures. This can for example be one texture with a list of data values or multiple textures. As an alternative, shader storage buffer objects could be used instead of the textures.

Please note that texture filtering needs to be disabled when accessing the grid texture via `texture` calls in GLSL. Otherwise, one has to make sure that the texels are only sampled at their exact centers to prevent interpolation. The recommended alternative is to use `texelfetch` instead for sampling the texture with non-normalized texture coordinates, i.e., grid indices, and without filtering.

As mentioned above, one of the drawbacks of the uniform grid is its constant cell size, which is not adaptive. But when embedding the atomic structure of a protein into a uniform grid this fact is not of high relevance. The spatial structure of a protein is typically very dense as well as space-efficient. Thus, a large number of grid cells will be occupied. This is of particular interest if we want to visualize for example the intracellular or extracellular environments which are filled with millions of proteins. The challenge here is to maintain the details on the nanometer scale for atoms while at the same time covering tens or even hundreds of micrometers between cells. This difference of at least three orders of magnitude makes this problem quite challenging. In particular, at these length scales we are typically also interested in a longer time scale, i.e., the atom movements within a protein are less noticeable while the diffusion and transportation of the proteins themselves in space becomes more dominant.

One might argue to enclose the entire scene in a single uniform grid instead of the individual proteins. First, such a grid would need to include billions of atoms, data which need to be stored and transferred to the GPU. Of course, more advanced methods like Octrees or kd-trees could be used for spatial subdivision of the cellular environment. But these methods are computationally expensive and require updating since we are dealing with a dynamic environment where proteins are moving around. Secondly, instancing proteins would no longer be possible. As we will see it is sufficient to apply uniform grids only to proteins for interactive rendering. In the following we will detail the concepts for rendering such a cellular environment with atomistic detail as presented earlier by Lindow et al. [2012] and Falk et al. [2013b].

5.1.1   TEMPLATE-BASED INSTANCING

Depicting the intracellular space with atomistic detail can help us gain a better understanding of cellular dimensions and complexity. The biggest challenge at this scale is the sheer number of proteins and, thus, the even higher number of atoms. Lindow et al. [2012] make therefore the following assumptions for visualizing molecular structures on the cellular scale. First, the scene contains billions of atoms but the number of individual molecule types is low, i.e., a high abundance of the same proteins. Secondly, molecules are considered to be rigid. Although not entirely correct, this assumption is based on the large differences on the time scale between atom movements and protein diffusion processes. Hence, the molecular structure needs to be stored only once per molecule type, thereby reducing memory requirements dramatically. Furthermore, we can employ an adapted version of glyph ray casting for the rendering of instances of atomic structures providing results of high quality.

The underlying idea is to embed each kind of molecular structure, i.e., protein, into its own uniform grid. This grid serves as acceleration structure and storage for the densely packed space of the protein. Instantiating this grid during rendering then triggers the ray casting within the grid, resulting in a visualization of the individual atoms. Since all the necessary information is stored in uniform grids residing in GPU memory, the overall number of draw calls corresponds to the number of molecule instances, i.e., several orders of magnitude less than the total amount of atoms. A combination of the tesselation shader and the geometry shader could further reduce the draw calls to a single instanced draw call for each molecule type.

The uniform grids represent lookup tables for the individual molecule instances. In a pre-processing step, the grid is filled by embedding all atoms of a molecule. The information on the atomic structures themselves is obtained from the protein data bank (PDB) [Berman et al., 2000]. The PDB data files provide information on the location and type of atoms as well as conformal information. In some cases, however, the file might only contain a part of the protein, which has to be combined with other protein parts. Structural elements like the cytoskeleton or virus capsids are composed of a larger number of proteins. For example, microtubules which are a part of the cytoskeleton consist of α- and β-tubulin dimers (PDB-ID: 1TUB) that are arranged in a circular way. For efficiency reasons, template grids for different lengths are generated and instantiated separately. This allows us to render elongated structures with varying lengths like filaments of the cytoskeleton.

For embedding the molecular structure in a uniform grid there exist basically two strategies. The first approach stores the information of each atom only once and avoids data duplication. That is, only the grid cell corresponding to the atom position refers to the atom. When testing for intersections, it is therefore necessary to test all neighboring cells as well. For more details on this strategy we refer the reader to the work by Lindow et al. [2012]. In the second approach, a grid cell refers to all atoms which intersect with this cell leading to potential data duplication. If the size of the grid cells is chosen with respect to the atom size the duplication is kept to a minimum. We recommend using a grid cell size of about $4\,\text{Å}$, i.e., $0.4\,\text{nm}$. Thus, it is ensured that a single atom is contained by at most eight grid cells [Falk et al., 2013b, Lindow et al., 2012]. Besides the slightly higher memory consumption, the advantage of this approach is the ease of finding potential ray-atom intersections within a grid cell.

5.1.2   ALGORITHM

Since the molecules do not undergo any internal deformations according to our assumptions (cf. Section 5.1.1), a single molecular template is sufficient per protein type and, thus, the uniform grids containing the atomic structure have to be transferred to the GPU memory only once. During rendering, multiple instances of the same molecule are drawn. This allows keeping the amount of data which has to be transferred for each rendering pass from the CPU to the GPU

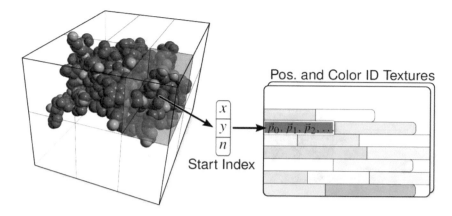

Figure 5.1: A protein (insulin, PDB-ID: 1RWE) is embedded into a uniform grid, where each grid cell refers to a consecutive list of atoms intersecting with this cell. The additional information, i.e., atom position, radius, and coloring, is stored in 2D textures.

is very low. For each rendered molecule, only a translation and a rotation transformation have to be transferred, which can either be stored in a single matrix or in a quaternion and an additional translation vector. In the vertex shader, the quaternion rotation must then be converted back into a matrix. This computational overhead is compensated by the fact that less data has to be transferred to the GPU, i.e., only seven values for a quaternion and translation compared to 12 values for a matrix.

The atoms of a template molecule are embedded into a uniform grid (see Figure 5.1). A box-sphere intersection test is used to determine all cells of the grid with which an atom intersects. The grid data structure proposed by Lagae and Dutré [2008] is used to store the data of the uniform grid. Each cell of the grid is mapped onto one voxel of a 3D texture. The atomic data, i.e., atom position, radius, and color IDs, are stored in two additional 2D textures. The first texture contains the atom position and its radius represented in 32-bit floating point values. The color identifiers, e.g., atom type, chain ID, or strand ID, are stored in the second texture with up to four channels. This additional data is accessed via a 2D index (x, y) stored in the 3D grid texture. In addition to the 2D index, each voxel also contains the number of atoms n in this cell. Imprecision in indexing can be avoided by using a 16-bit integer format for the 3D texture, thus allowing precise addressing of textures up to 32k × 32k texels.

Rendering
To initiate the ray casting of the atoms, only the grid's bounding box of each instance is drawn. By rendering only the back faces of the bounding box and computing the corresponding position on the front side, the ambiguity between front and back faces is avoided. The same effect can,

Figure 5.2: Visualization of a cellular mesoscopic simulation including parts of the cytoskeleton and signaling proteins with atomistic detail at 8 fps.

of course, also be achieved by rendering only the front faces and reconstructing the back sides. The fragment shader computes the view ray through each fragment covered by the bounding box and traverses the grid cells front to back. The regular depth test is used to maintain the correct depth order between protein instances. To avoid unnecessary shading calculations during ray casting we employ deferred shading (see Chapter 7 for details). Our shader therefore outputs material or color ID, the corresponding normal, and the scene depth. In addition, this approach also lends itself to the application of object space ambient occlusion as described in Section 7.2 since we can also store the ambient occlusion factor, which is sampled at the position of the ray-sphere intersection. In Figure 5.2, a close-up of the interior of a eukaryotic cell is shown. The data originates from a mesoscopic simulation and includes signaling proteins and filaments of the cytoskeleton. The entire scene contains about 25 billion atoms and renders at $3 - 15$ fps at 1920×1200 on a NVIDIA GTX 580.

Ray-grid Cell Intersections

Figure 5.3 illustrates the grid traversal for a single ray. The individual ray-sphere intersections for the atoms are consequently computed per grid cell. As soon as at least one atom is hit, the grid traversal can be stopped. It is, however, important to compute all intersection in one grid cell to obtain the closest intersection since the atoms are not ordered along the ray direction. The corresponding GLSL shader code for the intersection test is shown in Listing 5.1.

Listing 5.1: Ray intersection test for glyphs embedded in a uniform grid cell

```
1  bool rayIntersection(in ivec3 voxel, in float tMax,
2                       in vec3 rayPos, in vec3 rayDir,
3                       out vec4 dstColorId, out vec3 dstNormal,
4                       out float dstDepth) {
5      ivec3 texIndex = texelFetch(gridSampler, voxel, 0).stp;
```

```
 6        float tMin = tMax; // keep track of distance to closest glyph
 7        // check all glyphs inside this voxel
 8        for (int i=0; i<texIndex.z; ++i) {
 9            ivec2 texCoord = ivec2(texIndex.x + i, texIndex.y);
10            // read atom info
11            vec4 atomPos = texelFetch(atomPosSampler, texCoord, 0);
12            vec4 atomColorId = texelFetch(atomDataSampler, texCoord, 0);
13            // perform ray-sphere intersection
14            // tMin and outputs are only written to if the
15            // intersection is closer than tMin
16            rayGlyphIntersection(rayPos, rayDir, atomPos.xyz, atomColorId,
17                                 tMin, dstColorId, dstNormal);
18        }
19    if (tMin < tMax) {
20        // found intersection
21        dstDepth = computeDepth(vec4((rayPos + tMin * rayDir), 1.0));
22        dstNormal = normalMatrix * normalize(dstNormal);
23        return true;
24    } else {
25        return false;
26    }
27 }
```

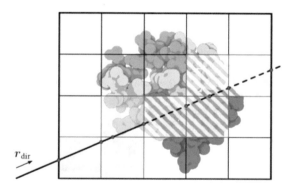

Figure 5.3: Traversal of uniform grid in combination with sphere ray casting of molecular structures.

First, the index is fetched from the 3D grid texture using the current voxel index and tMin is initialized with the maximum free ray length, i.e., the distance to the closest boundary of the grid cell. We then loop over all atoms in this cell, fetch their associated data, and perform a ray-sphere intersection for each atom. Here, tMin is an inout parameter, i.e., it can be changed inside the function. Please note that an intersection is only considered valid if its distance is less than the current tMin, which will in turn update tMin and the output parameters. To avoid unnecessary computations, the normal transformation and the depth calculation are performed after all atoms have been tested.

Ray-grid Traversal

To be able to determine the intersections between rays cast onto the bounding boxes of the molecule instances and the enclosed grid, we need to traverse the uniform grid. The traversal of the grid in the GLSL shader follows the idea of Amanatides and Woo [1987] for a fast and discrete ray-voxel traversal. Using integer arithmetic for the grid indices avoids grid sampling artifacts caused by floating point inaccuracies since the current grid is given accurately without need for division or the modulo operation. This is made possible by separating the spatial position of the ray casting from the voxel index. Listing 5.2 depicts the GLSL code of the grid traversal.

Listing 5.2: GLSL fragment shader segment for traversing a uniform grid using integer arithmetic as described by Amanatides and Woo [1987]

```
1   // compute initial voxel position
2   ivec3 voxelIndex = objectCoordToVoxel(rayPos); // (grid coordinates)
3   vec3 cellSize = bboxSize / gridDim; // (object space)
4   // offset of ray position from current voxel center
5   vec3 offset = ((voxelIndex + 0.5) * cellSize - rayPos);
6   // initial t from ray position to closest cell boundaries (object space)
7   vec3 t = (offset + 0.5 * sign(rayDir) * cellSize) / rayDir;
8   // delta inside grid cell from side to side along the ray (object space)
9   vec3 tDelta = cellSize / abs(rayDir);
10  ivec3 signDir = ivec3(sign(rayDir));
11
12  // grid traversal
13  // keep voxel index and ray position (object space) separate
14  float tCurrent = 0.0;
15  while (tCurrent < maxRayLength) {
16      ivec3 currentVoxelIndex = voxelIndex;
17      // 1) determine closest face of current cell,
18      // 2) set ray parameter tCurrent to closest face, tCurrent will be
19      //    used to determine valid hits inside the current grid cell
20      // 3) advance voxel index in that direction
21      if (t.x < t.y) {
22          if (t.x < t.z) { // x < y, x < z
23              tCurrent = t.x;
24              t.x += tDelta.x;
25              voxelIndex.x += signDir.x;
26          } else { // x < y, z < x
27              tCurrent = t.z;
28              t.z += tDelta.z;
29              voxelIndex.z += signDir.z;
30          }
31      } else {       // y < x
```

```
32          if (t.y < t.z) { // y < x, y < z
33              tCurrent = t.y;
34              t.y += tDelta.y;
35              voxelIndex.y += signDir.y;
36          } else { // z < y, y < x
37              tCurrent = t.z;
38              t.z += tDelta.z;
39              voxelIndex.z += signDir.z;
40          }
41      }
42      // test for intersections inside current grid cell in object space
43      // but consider only hits with t < tCurrent.
44      if (rayIntersection(currentVoxelIndex, tCurrent, rayPos, rayDir,
45                          colorId, normal, depth)) {
46          // found a valid hit point, set output color and depth
47          ...
48          return;
49      }
50  }
```

In the beginning, the respective voxel index of the ray entry position is determined. We then compute the minimum distance t along the ray from the entry position to the closest cell boundary for each axis (see Figure 5.4). By projecting the extent of a grid cell onto the ray we obtain Δt, the step length necessary to proceed to the next boundary. After these calculations we can start with the iterative grid traversal.

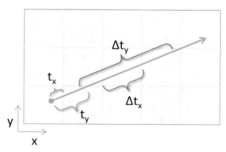

Figure 5.4: Integer-based grid traversal using step lengths $\Delta t_{x,y}$ and initial free distances $t_{x,y}$.

First, we save the current voxel index (line 16) and determine the closest boundary of the grid cell by finding the index i of the smallest component t (lines 21–41). The current step length $t_{current}$ of the ray is set to t_i and is used to limit intersection tests to the current grid cell and can, thus, be considered an upper limit. To proceed to the next grid cell we need to cross the selected boundary. Thus, the respective component i of the voxel index is advanced by one and Δt_i is added to t_i (lines 25, 29, 35, and 39, respectively). Then the ray intersection tests for the current grid cell are performed (line 44) using the saved voxel index and $t_{current}$. The main loop is executed as long as the ray does not exit the bounding box or a valid hit has been found.

Optimizations

Due to the large differences in the spatial domain, many atoms will cover at most one pixel of the final image when visualized at the cellular scale. This is also true for atoms which are located some distance from the camera. The low pixel coverage of atoms can be exploited to avoid the costly grid traversal. In particular, our uniform grid provides three inherent hierarchy levels. First, the bounding box of the templated instance. Secondly, the grid cells and whether they are occupied or empty. And last, the atom level with ray-sphere intersections.

Before the ray-intersection tests take place in a grid cell, the voxel size is compared with the pixel size corresponding to the entry point of the ray (Figure 5.5, left). The pixel size is obtained for the respective depth with the intercept theorem from the viewport height in pixels and the field of view. If the pixel is larger than the voxel, i.e., the voxel covers only a part of the final pixel, the ray-sphere intersections can be omitted because individual spheres will not be discernible. Instead of the intersection test, only one texture lookup is made to determine whether the voxel is empty or not, resulting in a binary voxelization. As the grid size was chosen with respect to the atom sizes, and the voxel contains at least one atom, the chance of one atom being hit by the ray is high and, hence, the grid traversal can be stopped for non-empty voxels. In this particular case, a special coloring scheme has to be employed since no intersection tests are performed: using the color of the first atom entry has the advantage that coloring according to chain, strand, or instance ID is still possible and also visible. Other possible color schemes could apply a precomputed averaged color per grid cells or just a single color. Similarly, we use the inverse view direction as normal, thereby assuming that we hit the atom in its center.

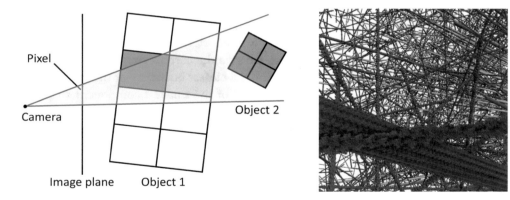

Figure 5.5: The hierarchical optimization for the grid-based ray casting provides full detail (blue) in the foreground where all ray-sphere intersections are performed. If the projected pixel size is larger than a grid cell, only filled grid cells (yellow) are considered. Molecule bounding boxes covering only parts of a pixel are drawn in red.

The approach is not only applicable to grid cells but also to the entire bounding box of the protein. This will prevent molecules from becoming invisible due to sizes smaller than a pixel. The combination of the regular grid traversal and the consideration of pixel sizes for both molecular bounding box and individual grid cell, results in a hierarchical ray casting and yields a performance boost. In Figure 5.5, the different levels of detail are highlighted when hierarchical ray casting is applied. Since the normals of far-away atoms are approximated during deferred shading, the binary voxelization of the grid does not show up in the final image.

Triangle Meshes

The grid-based approach described above relies on spherical glyphs, i.e., implicit geometry. Sometimes, however, one might be interested in rendering arbitrary surfaces like the solvent-accessible surface or the solvent-excluded surface of proteins. Following up on the idea of using instances of proteins, we can use the same approach to store and render triangle meshes. This is also a necessity for scenes containing thousands of proteins, as a surface representation of a single protein might already consist of hundreds of thousands of triangles, thus quickly exceeding even the capabilities of current high-end GPUs.

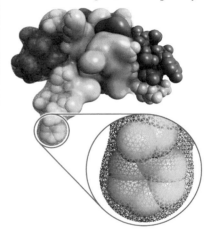

The basic approach is the same as for the spherical atoms. The triangles are embedded into the cells of the uniform grid by performing triangle-box intersections [Akenine-Möller, 2001]. The vertices of each triangle are then stored in the additional 2D textures. To avoid indirections in the shader, we store each triangle explicitly. Thus, the memory footprint is increased which, however, results in a performance gain. Eventually, the ray-sphere intersection code needs to be replaced with ray-triangle intersection tests, e.g., the efficient check proposed by Möller and Trumbore [1997].

Figure 5.6 depicts an example of triangulated solvent-excluded surface in combination with the atomic protein structure of insulin. Using this approach, the frame rate is about 6 fps for a scene containing 500 instances and a total of 110 million triangles. The performance is inferior to glyph ray casting which is due to the more expensive intersection tests and the sampling of the vertices. For more details on the performance results, we refer to Falk et al. [2013b].

Figure 5.6: Extending the grid-based glyph ray casting to render triangulated meshes, like the solvent-excluded surface, besides the atomic structure (insulin, PDB-ID: 1RWE).

5.2 HIERARCHICAL DATA STRUCTURES

The uniform grid in combination with the aforementioned optimization could be considered a hierarchical data structure. Although, the data structure is inherent with the grid and requires little overhead, it results in a rather crude simplification and approximation of the data for far-away proteins and atoms. In particular, the transition between the three levels of the hierarchy are

Figure 5.7: Semantic level of detail for 10^6 molecules. The abstraction is depending on the camera distance. The farther away the more atoms are removed from each protein and the remaining atoms are scaled up to maintain the appearance. From [Le Muzic et al., 2014].

non-smooth. What we want to have instead is a smooth abstraction which is explicitly defined and, thus, provides us with more control and flexibility.

In many scenarios we would therefore like to apply a semantic abstraction to retain certain features. For rendering proteins and other molecules the criterion for simplification might be to maintain the overall visual appearance, i.e., the protein surface which for example is corresponding to the van der Waals-surface of its atoms. The technique by Le Muzic et al. [2014], discussed in Section 5.2.1, achieves this by removing particular non-contributing atoms.

A different approach is to utilize the hierarchical data structure itself. By computing the hierarchy based on the underlying point data, e.g., by clustering, we can use this information both for data abstraction and reducing the memory required to store the data. The concept of coordinate quantization in combination with a hierarchy [Hopf and Ertl, 2003] is described in Section 5.2.2.

5.2.1   IMPLICIT HIERARCHY

In 2014, Le Muzic et al. [2014] proposed a continuous level of detail approach for rendering scenes full of proteins without explicitly storing the hierarchical data. With this approach they were able to render 1 million molecules at 30 fps. Figure 5.7 illustrates how the individual atoms become discernible when zooming in while the general appearance of the proteins remains without major changes.

Semantic Abstraction

The basic idea of this approach is to simplify proteins where it hardly shows, i.e., to reduce the number of atoms hidden away inside the protein. The abstraction, or simplification, is solely based on the distance toward the camera. The larger the distance the more atoms will be removed. To fill potentially occuring gaps, the remaining proteins are enlarged. Figure 5.8 illustrates this simplification for one molecule.

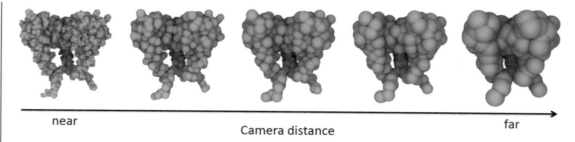

Figure 5.8: Semantic abstraction for molecular level-of-detail. The number of skipped atoms increases with the camera distance. Courtesy of Mathieu Le Muzic.

Near and far boundary values define at what distance the simplication will start and when it will reach its maximum. In addition, we need to determine the number of atoms visible at both boundaries. At the lower bound all atoms of the protein should be visible, whereas only a fraction should remain at the upper boundary. A resonable value for the upper bound is, however, subject to the overall visual appearance and personal preference. Between these two limits the number of visible atoms is interpolated linearly, thus ensuring a smooth transition. In case the distance exceeds the upper boundary, more atoms will be skipped and only every n-th atom is rendered.

To preserve the structure of the protein and avoid popping artifacts when removing multiple atoms, the authors suggest the following strategy. The atoms are sorted according to their increasing distance from the center of the molecule and stored in that order within a texture. Thus, n interior atoms can easily be skipped. The visual coherence between subsequent frames and also for varying camera distances is ensured since the order of the skipped atoms remains constant. The radii of the remaining atoms are scaled up by scaling factor which is obtained by linear interpolation between the near and far boundary. Despite the fact that this simple abstraction strategy is already quite effective, one could think of other importance critera. One example would be to consider not only the distance between atoms and the molecule's center, but to incorporate the structural information of the protein, e.g., different chains, as well.

This approach does not require a specialized data structure nor does it store any additional data besides the ordering of the atom, which could be obtained in a preprocessing step. Nonetheless, it can be considered an implicit hierarchical data structure for particle data, or in this case, molecular structures.

Rendering

Since the number of atoms per scene can easily exceed one billion atoms, i.e., assuming one million molecules with on average several hundred to thousands of atoms, we cannot solely rely on the standard GPU pipeline. The load on vertex and fragment processing would be too high for simply rendering vertices to trigger the glyph ray casting for all atoms.

Le Muzic et al. [2014] utilize the fact that the scene contains many similar molecules, as assumed above in Section 5.1 for applying uniform grids. The rendering of each molecule is triggered by drawing a single vertex along with the molecule's position and orientation. The structural data of the molecule is then accessed and recreated in the tesselation and geometry shaders. This bears high resemblance with the approach by Lampe et al. [2007], where proteins are composed of individual amino acids which are in turn used to instantiate the atoms with the geometry shader.

The combination of tesselation shader and geometry shader is necessary since the output of the tesselation shader is limited to 4,096 vertices due to limitations in current hardware. To instantiate more atoms, the geometry shader can be run on the output of the tesselation and, thus, generate another 64 atoms per vertex, totaling in at most 262,144 atoms per rendered vertex. The tesselation control shader uses isolines as patch primitives and the outer tesselation levels, i.e., gl_TessLevelOuter, are set to $\sqrt{\#\text{atoms}}$ each. The calculation of the individual atom IDs in the tesselation shader is given in Listing 5.3.

Listing 5.3: Calculation of atom IDs in the tesselation shader

```
1  ID = gl_TessCoord.x * gl_TessLevelOuter[0]
2       + gl_TessCoord.y * gl_TessLevelOuter[0] * gl_TessLevelOuter[1];
```

Given the atom ID, we can access the relative atom position and its radius from a texture. The geometry shader is then again used to create the glyphs for all atoms to trigger the sphere ray casting. Additional view frustum culling is performed in vertex shader to prevent molecules from being instantiated in case they are not visible.

The semantic abstraction is easily incorporated into this workflow. First the level of abstraction, i.e., the number of skipped atoms n, is determined based on the distance to the camera (see above). Since the atoms are sorted according to relevance, we only need to instanciate #atoms $- n$ atoms and apply the respective scaling factor. The result is an interactive visualization of the cellular environment with a continuous abstraction. Figure 5.9 depicts the smooth transition. For illustrative reasons, the abstraction level varies from left to right and is independent of the camera distance.

5.2.2 POSITION COORDINATE QUANTIZATION

The molecular datasets we have been dealing with so far in this chapter featured a high repetitiveness in the data. This has been exploited to store only a fraction of the atoms on the GPU and rely on instancing. However, this assumption does not hold in general. With the ever increasing size in particle datasets, different methods and data structures are required. We now discuss one possible solution supporting adaptive rendering and data compression through coordinate quantization, both enabled by a hierarchical data structure. The original work dates back more than ten years and was presented by Hopf and Ertl [2003] using it to visualize the results of n-body simulations performed by the VIRGO supercomputing consortium.

Figure 5.9: Continuous level of abstraction from coarse (left) to individual atoms (right). Courtesy of Mathieu Le Muzic.

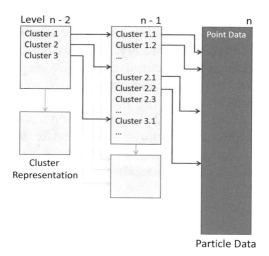

Figure 5.10: Cluster hierarchy of particle dataset with hierarchy information stored separately for improved rendering performance. Based on [Hopf and Ertl, 2003].

Hopf and Ertl use multiple PCA-splits to create a cluster hierarchy from particle data. The proposed data structure (see Figure 5.10) reflects this hierarchy, i.e., a cluster at level $n - 1$ links to its subclusters at level n as well as to a cluster representative. All representatives of the same level are stored next to each other in a continuous block. Hence, we can render all points of a selected cluster subtree at a specific level with a single draw call. The cluster hierarchy is completely decoupled from the actual data of the particles and the cluster representatives. In Figure 5.11, the different levels of the cluster hierarchy of a VIRGO n-body simulation are shown. The cluster representatives are scaled to maintain a brightness similar to all contained particles.

Figure 5.11: Visualization of the cluster hierarchy of a VIRGO n-body simulation from coarsest to finest level (left to right). From [Hopf and Ertl, 2003].

In addition, the cluster hierarchy enables us to use relative coordinates where the corresponding cluster centroid serves as a reference. Coordinate quantization can then be applied to the relative positions since the contents of a cluster are spatially contiguous by definition and, thus, less precision is required.

The results for another VIRGO simulation run are depicted in Figure 5.12. An adaptive rendering is shown on the left with the reference image on the right. The dataset contains 16 million particles requiring 256 MiB of GPU memory for 3D positions and corresponding radii, both stored as 32 bit floats. In this example, the quantization of particle positions reduced the data size by almost 40%. The necessary information to save the hierarchy requires an additional 12% for six levels and using bytes as underlying data type, which is sufficient to end up with floating point precision for the particle positions. To conclude, applying coordinate quantization to particle data will in most cases result in saving more memory than required for the hierarchy itself. In addition, the hierarchy lends itself to adaptive rendering for particle datasets with the following restrictions: either, one can afford to hide the precision errors in subpixels or the data already contains a coarse, natural representation.

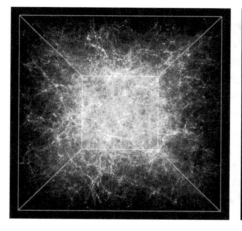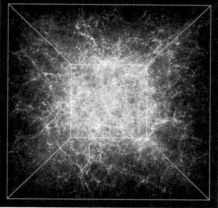

Figure 5.12: Adaptive point-based rendering with a screen-space error of 2 pixel (left) and the original data (right, 16.8 million points) The colors represent the velocities of the individual galaxies. From [Hopf and Ertl, 2003].

CHAPTER 6

Efficient Nearest Neighbor Search on the GPU

Finding the neighboring particles of a particle is a common problem for many computational tasks. Mostly, a *fixed-radius nearest neighbor search* is required. That is, all neighboring particles that are within a certain distance have to be found (see Figure 6.1). A common application is the detection of interacting particles. This is not only necessary for visualization, but also for other areas such as simulation, for example molecular dynamics (see, e.g., Frenkel and Smit [2001]) or smoothed particle hydrodynamics (see, e.g., Gingold and Monaghan [1977] or Monaghan [2012]). In biochemistry, for example, two atoms are considered to be covalently bonded if their distance equals the sum of their covalent radii. Consequently, covalent bonds can also be estimated based on the van-der-Waals radius: if two atoms are so close to each other that their van-der-Waals radii overlap to a certain amount, they have to be covalently bonded. That is, covalent bonds do not have to be stored explicitly, they can

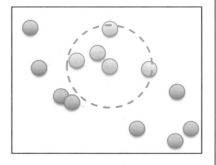

Figure 6.1: Illustration of a fixed-radius neighbor search for the green particle. The search area is defined by the green dashed circle and its radius; the neighboring particles are colored turquois.

be identified by checking the distance to all neighboring atoms. Since the van-der-Waals radius of each element is fixed, an upper bound can be defined for the neighbor search regardless of the element. Only the atoms within this fixed search radius have to be checked to find all bonded atoms. In visualization, finding covalent bonds is essential for the commonly used ball-and-stick representation depicted in Figure 6.2, which shows these bonds explicitly. Other types of bonds, like hydrogen bonds or disulfide bridges, require additional criteria to be met besides the distance. Examples for other applications that require a fixed-radius neighbor search include clustering or the extraction of surfaces from particles (see Chapter 8).

Visualization properties might be based on the neighborhood of a particle. For dynamic data, however, the particle neighbors can change for each frame. Therefore, a highly efficient neighbor search is needed in order to be able to interactively deal with dynamic particle data. A naïve approach would be to compute the distance to all other particles for each particle, which would induce a time complexity of $\mathcal{O}(n^2)$. Since this would clearly be too costly for

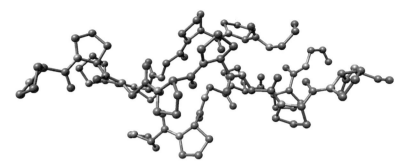

Figure 6.2: Ball-and-stick rendering of a small protein (PDB ID: 1A3I). The covalent bonds are shown as cylinders that connect the spheres representing the atoms. The bonds and atoms are colored according to the respective chemical elements.

an interactive computation of large numbers of particles, an efficient data structure for spatial subdivision has to be used, which allows us to retrieve neighboring particles. This data structure has to be either updated in real time or recomputed on the fly if particles move.

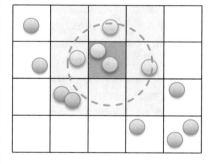

Since, in the general case, all particles move all the time, it is typically simpler and more efficient to recompute the entire data structure instead of performing partial updates. In order to attain real-time performance for the neighbor search, the parallel computing power of the GPU can be leveraged. Employing the GPU is especially beneficial since the particle positions have to be transferred to the GPU anyway for rendering, i.e., the neighbor search will not introduce any overhead with respect to data transfer between CPU and GPU. Furthermore, the results of the neighbor search will be available directly on the GPU for further parallel processing.

Figure 6.3: Nearest neighbor search on a grid. Not only particles within the grid cell containing the green particle (dark gray) but all grid cells within the search radius (light gray) have to be checked for neighbors.

One of the most simple approaches to spatial subdivision is a uniform regular grid. The spatial domain that encompasses the particles is partitioned by the grid cells. Each particle is stored in the respective grid cell that contains its center. During neighbor search, only particles that are located in grid cells that are within the search radius have to be tested. A grid data structure has the benefits that it is relatively storage-efficient, maps well to the GPU, and can be constructed efficiently on the GPU.

For sparse data sets with large empty regions, more complex data structures like octrees [Meagher, 1980] or k-d trees [Bentley, 1975] can be beneficial since they allow for empty space skipping. Especially for interactive GPU-based ray tracing, k-d trees are typically used [Horn et al., 2007], and their efficient implementation on the GPU is still an active area of research (e.g., Gieseke et al. [2014]). These data structures, however, also require more computations for constructing them as well as traversing them. Consequently, their use has to be considered carefully based on the application area. For molecular data or particle-based fluids, for example, the particle density is based on the laws of physics. That is, particles will not be packed arbitrarily densely. Consequently, only a certain amount of particles will be in the neighborhood of each particle. Empty regions can be minimized by using a tight-fitting bounding box of the particles for the grid. Therefore, a grid-based spatial data structure is usually preferable in these application areas.

In the following, the GPU implementation of a uniform grid-based spatial data structure is explained. The implementation was developed by Green [2012] for a CUDA particle simulation, which is part of the *NVIDIA CUDA Toolkit*.[1] This data structure can be used to accelerate the fixed-radius nearest neighbor search efficiently. The sample implementation uses CUDA; however, it could also be implemented analogously using any other GPGPU API (e.g., OpenCL or Compute Shaders).

The first step in the algorithmic pipeline is to sort the particles into the grid. Since multiple particles can be located within the same grid cell, a parallelized implementation has to avoid write conflicts explicitly. Otherwise, parallel threads can overwrite the memory location concurrently written by another thread. There are typically two solutions to avoid this issue: Atomic operations, which are guaranteed to operate exclusively on a memory location, or designing the algorithm so that it does not create write conflicts. While the first solution is usually simpler to implement, the second one can result in lower execution times, since no thread will be stalled by concurrent threads writing the same memory location. The grid construction of Green follows the second principle using a spatial hashing to compute the acceleration grid. For each particle, a hash value is computed based on the index of the corresponding grid cell that contains the particle center. The grid cell index (x_i, y_i, z_i) for particle i located at position (p_x, p_y, p_z) is computed as follows:

$$x_i = \left\lfloor \frac{p_x - min_x}{g_w} \right\rfloor \tag{6.1}$$

where x_i is the x-index of the grid cell, p_x is the x-coordinate of particle position i, min_x is the minimum value of the grid bounding box on the x-axis, and g_w is the width of a grid cell. The indices for y_i and z_i are computed analogously using the height and depth of the grid cells. Using this grid cell index, the hash value of particle i can be computed as the linearized cell index in the acceleration grid:

$$h_i = g_x \cdot g_y \cdot z_i + g_x \cdot y_i + x_i \tag{6.2}$$

where h_i is the hash value of particle i, g_x and g_y are the number of grid cells along the x-axis or y-axis, respectively, and (x_i, y_i, z_i) is the index of the grid cell that contains the particle i. After uploading the particle positions to GPU memory, the hash values are computed in parallel for

---

[1]CUDA Toolkit | NVIDIA Developer: https://developer.nvidia.com/cuda-toolkit (last accessed 08/19/2016).

each particle and written to a separate array, i.e., no write conflicts can occur at this stage. In the next step, the particles are sorted according to their hash value. This can be achieved using for example the parallel sorting methods provided by the Thrust library, which is part of the Nvidia CUDA Toolkit and offers the function `thrust::sort_by_key`. Since the hash depends on the grid cell index, the particles are sorted according to the grid cell they are located in. Figure 6.4 shows a simple example in 2D.

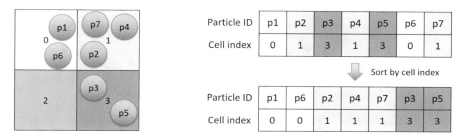

Figure 6.4: 2D schematic of the particle sorting step for seven particles $p1, \ldots, p7$ and a grid of $2 \times 2 = 4$ grid cells. First, the cell index for all particles is computed (top right). Afterward, the particles are sorted according to their cell index (bottom right).

Each grid cell can contain an arbitrary number of particles; therefore, the start and end index within the sorted particle list has to be found for each grid cell. In the final step, the hash of each particle i is compared to the hash of its predecessor $i - 1$. If the hash values differ, the predecessor is located in another grid cell. That is, the current particle i is the first particle in its grid cell j, whereas the previous particle $i - 1$ is the last one in its grid cell k. Conversely, cell end index for grid cell j is i and cell start index for grid cell k is $i - 1$. These values can be written to the respective arrays since no write conflicts can occur (the start and end indices of each grid cell are each written only once). A schematic example is shown in Figure 6.5. Listing 6.1 shows the CUDA source code that writes the start and end indices. Note that this function reads two hash values (the hash of the current particle and the one of the previous particle), i.e., each particle. The original version of Green uses shared memory to avoid these redundant reads. Here, each thread

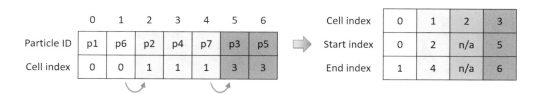

Figure 6.5: 2D schematic of the start and end indices of each grid cell within the sorted array of particles. Note that no values are written for the empty grid cell (2).

first writes only the hash of its particle index to an array in shared memory and synchronizes the threads afterward. The rationale for this is that reading the neighboring hash from shared memory is faster than reading it from global memory.

Listing 6.1: CUDA kernel that writes both start index and the end index of all grid cells. The indices point to the particle list that is sorted by the hash value (i.e., the grid cell each particle is located in). The number of threads has to be equal to or larger than the number of particles.

```
// Parameters:
// - nPart:          Number of particles
// - partHash:       Array of particle hashes (= grid cell indices)
// - cellStart:      Start indices of the grid cells (output)
// --> cellStart[m]: index of first particle that is in cell m
// - cellEnd:        End indices of the grid cells (output)
// --> cellEnd[m]:   index of first particle that is not in cell m
//                   anymore (i.e., x < y)
__global__ void makeCellStartEndList(uint nPart, const uint *partHash,
                                     uint *cellStart, uint *cellEnd) {
    // compute the index of the current thread
    unsigned int index = (blockIdx.x * blockDim.x) + threadIdx.x;
    // do nothing if current thread index exceeds number of particles
    if (index < nPart) {
        // read hash value of the current particle
        uint currentHash = partHash[index];
        // read hash value of the previous particle
        uint previousHash = partHash[index-1];
        // Since the particles are sorted, if the current particle has
        // a different cell hash than its predecessor, it is the first
        // particle in its cell, and its index marks the end of the
        // previous cell.
        if (index == 0 || currentHash != previousHash) {
            // set the start index of the current cell
            cellStart[currentHash] = index;
            if (index > 0)
                // set the end index of the previous cell
                cellEnd[previousHash] = index;
        }
        if (index == nPart - 1) {
            // set the end index of the last grid cell
            cellEnd[currentHash] = index + 1;
        }
    }
}
```

In order to find all neighboring particles to a position $p \in \mathbb{R}^3$ within a radius r, the grid cell index $j = (j_x, j_y, j_z)$ has to be computed. Next, the number of grid cells that are within the search radius has to be determined for each axis:

$$n_x = \left\lceil \frac{r}{g_w} \right\rceil, \quad n_y = \left\lceil \frac{r}{g_h} \right\rceil, \quad n_z = \left\lceil \frac{r}{g_d} \right\rceil, \tag{6.3}$$

where n_x, n_y, and n_z are the number of grid cells in the direction of each axis, and g_w, g_h, and g_d are the width, height, and depth of a grid cell. The neighboring particles will be in the block of cells between $(j_x - n_x, j_y - n_y, j_z - n_z)$ and $(j_x + n_x, j_y + n_y, j_z + n_z)$. In order to find only the particles that are within radius r of p, the distance between p and each particle located in one of these grid cells has to be computed. This step can be skipped if the whole grid cell is within the search radius r, since in this case, all particles within this grid cell will also be within r.

The fixed-radius nearest neighbor search described above using a GPU-accelerated parallel sorting is a fast, general solution since it avoids write conflicts. However, Hoetzlein [2014] recently found that on newer Nvidia GPU architectures the use of *atomic* operations does not always cause a slowdown. The idea of their *counting sort* algorithm is to count the number of particles in each grid cell while computing and writing the hash values. This is achieved via an atomic counter per grid cell that is incremented for each particle. After computing the hashes and counting the particles per grid cell, a parallel prefix sum is used to sum up the particle count (e.g., using the `thrust::scan_exclusive` function provided by the Thrust library). This also provides the offsets for each grid cell, since it gives the total number of particles in all previous grid cells. Using these offsets, each particle can now be written to the result array so that all particles that are located in the same grid cell are next to each other and sorted by the grid cell index, just as in the implementation described above.

CHAPTER 7
Improved Visual Quality

Applying the techniques presented earlier in Chapters 3 and 4, we can assume that rendering speed issues are resolved. We can visualize data as fast as possible, i.e., we can interactively view data sets with billions of particles on a single desktop machine. However, when viewing such large data sets problems with the visual quality occur, which are mainly related to aliasing. Figure 7.1 shows such an example. Obviously, these aliasing effects are not only a visual nuisance, but actually hinder perception of the data. This especially affects the understanding of shape and depth of structures formed by the individual particles.

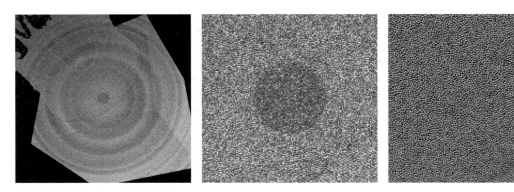

Figure 7.1: Rendering of a data set showing all particles (left), an enlarged close-up of the left image showing individual pixels (center), the same close-up rendered directly with glyphs.

The aliasing results from the fact that the individual particles only influence single pixels in the final image. Thus, only a single ray is cast for the particle which hits the glyphs surface almost arbitrarily, resulting in massive undersampling artifacts. Of course, aliasing affects all visual attributes of the glyphs, like color. But the most problematic attribute is the surface normal used for lighting computation. The resulting brightness, especially if the material model contains specular highlights, varies extremely between pixels, even if the corresponding particles form a rather continuous structure, like the surface of a solid object (cf. Figure 7.1). To address this issue, the first concept in this chapter (cf. Section 7.1) introduces deferred shading with image space normal estimation. Since lighting in general is the most pressing issue here, Section 7.2 presents some techniques on ambient occlusion lighting of particle data.

7.1   DEFERRED SHADING

In general, deferred shading is an acceleration technique. It allows us to limit costly operations to only the visible fragments, mostly used for lighting calculations. The scene is rendered into a framebuffer object, often called *G-Buffer* (from geometry buffer), holding all surface information to evaluate the final color, e.g., position, normal vector, and base color. In case of optimized particle-based visualization the acceleration, however, is not possible anymore. The advanced culling strategies presented in Section 4.3 already reduce the overdraw in the scene to a minimum. A further speed up cannot be expected as the additional image-space rendering pass will introduce a fixed overhead. Details on this discussion can be found in the corresponding publication [Grottel et al., 2010a]. But deferred shading as image-space pass provides the opportunity for post-rendering filtering, and even data and signal reconstruction. And this is very beneficial to eliminate visual sampling artifacts (cf. Figure 7.1, right).

The primary visual artifact is aliasing due to undersampling if particle sizes drop below pixel sizes. In scientific data sets the particle colors usually encode spread values, i.e., values that do not vary strongly between neighboring particles. For example, local density or energy are always similar within local regions. One exception to this rule would be encoding the particle type as color. In the case of mixtures, directly neighboring particles might have extremely different colors. Then, however, the individual colors should not be of high relevance when zooming out. Only the presence of different colors, and maybe a dominant color, are interesting, both information which are not really hindered by aliasing. The fragment attribute suffering most severely from the undersampling is the surface normal, as mentioned above.

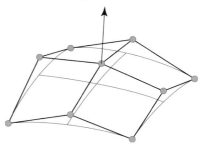

Figure 7.2: Schematic depiction of the image-space normal estimation. Nine fragment positions (blue) define the control mesh (black) of an approximating surface patch (red).

As soon as the image-space size of particles becomes smaller than individual pixels the glyphs' own surface information is no longer relevant. Hence, the whole costly ray casting can be omitted and simple splats into the geometry buffer suffice. Superstructures formed by the particles are still represented this way, following the concepts of point-based graphics. To (re-)construct the normal vectors required for the lighting computations these superstructures can be approximated in image-space. The G-buffer carries positional information in each fragment. Thus, to approximate the surface of the superstructure the local fragment neighborhood can be used.

Figure 7.2 shows a sketch of the approach. For evaluating the normal at each fragment, the eight direct neighboring fragments are also retrieved from the G-buffer. The positions of all nine fragments define a bi-quadratic patch approximating the superstructures' surface. The normal vector is then simply evaluated for this patch at the position of the original center fragment.

There are some caveats which need to be avoided. One obvious problem is that the surface patch can only exist under the assumption that all nine fragments form one closed surface. This is not the case, e.g., if background fragments are involved, for fragments outside the image viewport, and for fragments with strongly varying depth values. All these cases need to be addressed. A simple heuristic is to omit neighboring fragments if they meet one of these three conditions. While the first two are trivial to test, the variation in depth strongly depends on the rendered data. A working heuristic is to check the depth differences between the inner fragment d_0, the fragment on one side, e.g., fragment d_1, and on the opposite side d_{-1}.

$$d_1 - d_0 > \gamma(d_0 - d_{-1}) . \tag{7.1}$$

If Equation 7.1 does not hold true for a user-specified value of γ, then d_1 is assumed to be not part of the same superstructure. Instead of handling these cases with special cases of the patch evaluation, the data of the control mesh can simply be adjusted. The inner fragment position is always part of this patch. The control point representing an omitted fragment is moved toward the center point half of the distance. The rationale for that is to limit the patch to the spatial extend of the center fragments. Similarly for the depth value of this control point, the depth value of the center point is used. Assuming all eight neighboring fragments would be invalid, the patch would end up being aligned perpendicularly to the viewing direction and, thus, be flat. If only some neighboring fragments would be omitted, the patch would approximate them as best as possible. There are arguably more sophisticated heuristics for this problem, but we found this simple solution working quite well.

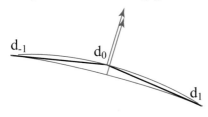

Figure 7.3: 1D example of the surface patch definition. Approximating patches (dashed blue line) effectively smooth the surface (black) compared to interpolating patches (dotted red line).

The actual patch definition still seems to hold some modeling freedom, as the patch can be defined approximating or interpolating. To define an interpolating patch the depth of the inner control points needs to be adjusted accordingly (cf. Figure 7.3). However, for the calculated normal at the center of the patch, this does not make any difference. For the sake of simplicity we assume that all fragments are valid and placed on a uniform grid. In the 1D case the curve then reads:

$$d(x) = (1 - x)^2 d_{-1} + 2(1 - x)x d_0 + x^2 d_1 . \tag{7.2}$$

To calculate the required normal vector at the center of the patch, we only need to evaluate the derivative of Equation 7.2 at $x = 0.5$. Note that this yields the well known central differences:

$$
\begin{aligned}
d'(x) &= 2(x - 1)d_{-1} - (2x - 1)d_0 + 2x d_1 , \\
d'(0.5) &= 2(0.5 - 1)d_{-1} - (2 \cdot 0.5 - 1)d_0 + 2 \cdot 0.5 d_1 , \\
&= -d_{-1} + d_1 .
\end{aligned}
$$

The actual value of d_0 does not influence the result. The patch normal vector, thus, simply results from central differences. Figure 7.4 compares renderings between the aliasing prone ray casting and the presented normal estimation.

Figure 7.4:   Rendering a laser ablation data set with 60 million particles. Left: aliasing artifacts result from undersampling during ray casting. Right: image-space estimated normal vectors clearly show emerging superstructures. From [Grottel et al., 2010a].

However, this normal vector estimation is only a heuristic. If the spherical particles are big enough the normal vectors resulting from ray casting are more precise, especially at the edges of the particles, and should, thus, be preferred. The glyph silhouette approximation performed in the vertex shader directly calculates the image-space size of the glyph. Based on this size the normal vector cannot only be chosen, but interpolated between the result from the ray casting and the result from the image-space estimation. For example, if the glyph size is larger than three pixels, use the ray casting normal. If the glyph size is one pixel or smaller, use the estimated normal. If the glyph size is between one and three pixels, blend linearly between the two normal vectors. Note that one should, of course, use floating-point precision for the image-space sizes here.

7.2   AMBIENT OCCLUSION

The previous section (Section 7.1) was based on the argument that the main reason for the visual artifacts are sampling artifacts primarily made visible by the lighting computations. Therefore, a valid approach to this issue is to change the lighting model as a whole. Several papers incorporate different computer graphics effects to increase the visual quality of particle-based visualization (e.g., Falk et al. [2013a], Tarini et al. [2006]). The techniques developed in interactive computer graphics are thus promising approaches. One simple but established approach is *Ambient Occlusion* (AO), approximating global illumination effects with a smooth darkening function. See the work by Grottel et al. [2012b] for a detailed introduction.

Ambient occlusion is based on local geometry properties of the scene, which are the features we want to emphasize. This method was first described by Zhukov et al. [1998]. They modeled the ambient lighting term in local lighting equations as radiating, non-absorbing, and completely

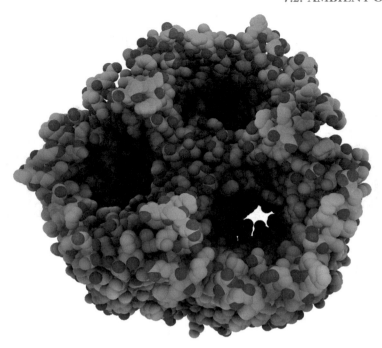

Figure 7.5: A protein data set (PDB ID: 1AF6 [Wang et al., 1997]) rendered with object-space ambient occlusion [Grottel et al., 2012b]. No local lighting operations are performed, only the ambient occlusion factors are multiplied with the particle base colors.

transparent gas equally distributed throughout the entire scene. This ambient lighting value is evaluated from the amount of the gas surrounding scene surfaces, which again depends on the distances of neighboring scene objects, figuratively speaking, how much potentially incoming light is blocked by surrounding geometry. While ambient occlusion approximations in image space are widely used in computer graphics, they are prone to create imperfect or even incorrect shadowing effects since only the visible parts of the scene are taken into account. Techniques that evaluate these ambient occlusion factors from scene information in object space are thus called *object-space ambient occlusion*.

The principle idea of ambient occlusion is to calculate a factor A_p for ambient light at each (visible) point p in the scene, which approximates the light distribution of a global illumination model. More precisely, the ambient occlusion factors simulate the lack of incident secondary light rays reaching point p, because their directions are blocked by nearby scene elements. The ambient occlusion factor A_p at point p is basically given by the distances of the occluding geometry in the visible hemisphere hS^2, while the distances are upper bound to only take a local neighborhood into account. These factors can then be multiplied with the results of a local lighting calculation to create the final image. But even only using the base color of the particles results in clearly visible superstructures (see Figure 7.5).

Equation 7.3 is a simplified form of the original equation from Zhukov et al. [1998] to calculate the ambient occlusion factor:

$$A_p = \iint\limits_{x \in hS^2} \rho(L(p,x)) \cos \alpha \; dx \; . \tag{7.3}$$

A_p results from the integral over the visible hemisphere hS^2 over the blocked incoming light energy $\rho(L)$ based on the distance to the nearest obstacle $L(p,x)$ in direction x. α is the angle between x and the surface normal n_p at p and, thus, $\cos \alpha = n_p \cdot x$ is a basic form factor for incoming light. For mesh-based geometry, $\bar{p}$ is selected to represent a patch and the resulting A_p is interpolated between patches to get a continuous result:

$$A_p = \sum_{\bar{p}_i \in P_p} \omega(p, \bar{p}_i) A_{\bar{p}_i} \; , \tag{7.4}$$

where P_p are the patches in proximity of p and $\omega(p, \bar{p}_i)$ is the interpolation weight factor for the value of patch $\bar{p}_i$ when evaluated at position p. $\omega(p, \bar{p}_i)$ basically denotes any interpolation, e.g., bi-linear factors or barycentric coordinates.

Equation 7.4 is only applicable because A_p is a continuous function over p. The integral over the hemisphere in Equation 7.3 basically represents the local neighborhood of p which changes only very little for near p. We therefore define an *occlusion geometry value* $O_{p,d}$ at position p in direction d based on Equation 7.3:

$$O_{p,d} = \iint\limits_{x \in hS_d^2} \rho(L(p,x)) d \cdot x \; dx \; , \tag{7.5}$$

with hS_d^2 describing the positive hemisphere based on the normalized direction vector d. Equation 7.4 can now be rewritten accordingly:

$$A_p = \sum_{\bar{p}_i \in P_p} \omega(p, \bar{p}_i) O_{\bar{p}_i, n_{\bar{p}_i}} \; , \tag{7.6}$$

with $n_{\bar{p}_i}$ being the normal vector of patch $\bar{p}_i$. The interpolation can now be performed on the *occlusion geometry values* instead of interpolating the ambient occlusion factors resulting in:

$$O_{p,d} = \sum_{\bar{p}_i \in P_p} \omega(p, \bar{p}_i) O_{\bar{p}_i, d} \; , \tag{7.7}$$

which yields $A_p = O_{p,n_p}$. The more dense the packing of particles is the less light rays are able to reach p from the direction x. Equation 7.3 can therefore be re-written:

$$A_p = \iint\limits_{x \in hS^2} D(p + \lambda x) n_p \cdot x \; dx \; , \tag{7.8}$$

with D being a density volume of particles and λ being a sampling distance from p. The size of the neighborhood used in this approach is implicitly defined by the discretization of the density volume, meaning that larger density voxel cells will result in larger neighborhood ranges being used in our approach. Thus, λ should be half the length of a side of a voxel cell. This value corresponds to $L_{\max}$ in Zhukov et al. [1998]. If we want to observe a reasonable neighborhood, we actually need rather large voxel cells, resulting in a coarse resolution density volume. From a performance point of view this is even more beneficial. Note that the built-in tri-linear interpolation of volume data provides a simple solution to ensure the required smoothness property of the data.

This coarse volume resolution allows for a further simplification: The integral over the hemisphere in Equation 7.8 will only fetch very few real data values, since the radius λ of the hemisphere is the diameter of only a single voxel cell. In addition, the geometry form factor $n_p \cdot x$ will reduce the influence of the values fetched for the border areas of the hemisphere. Thus, we can further approximate the *occlusion geometry value* by only fetching the most relevant, interpolated density value per patch:

$$A_p \approx D(p + \lambda n_p) \, . \tag{7.9}$$

To generate the required density volume, one can use a simple scatter approach:

$$V = \sum_{p \in V_{bounds}} \frac{4}{3} \pi r_p^3 \frac{1}{V_{size}^3} \, , \tag{7.10}$$

with V being a voxel value, $p \in V_{bounds}$ all particles being inside this voxel cell, r_p being the radius of a particle, and V_{size} being the length of one edge of the voxel cell. Equation 7.10 may raise several questions about its correctness:

First, the density might be overestimated if spheres overlap. The worst-case scenario is two spheres of the same size placed in exactly the same position, resulting in a volume twice as large as would be correct. This case, however, is not relevant for particle data from molecular dynamics when choosing the correct radius for the spheres. For example, using the Lennard-Jones radii on data from simulations using matching pair potential force fields, the spheres can only overlap very slightly if the simulation runs correctly.

Second, the closest sphere-packing needs to be addressed. A voxel cell should be considered completely full and opaque if the summed up volume of spheres reach $\approx 74\%$ of the voxel cells' volume. While this correctly models the aspect of the close-packing of spheres, it does not address the fact that a voxel cell might be completely opaque with far fewer spheres. Seen from one direction a voxel cell can be completely opaque, if there are just two layers of non-overlapping spheres perpendicular to this viewing direction. Obviously this cannot be addressed in general by only storing a density volume without any additional information. A simple solution is a user-defined factor g for a linear mapping of number of spheres to voxel cell opacity.

The third issue with Equation 7.10 is that the term $p \in V_{bounds}$ does not define what happens to spheres that are partially inside the voxel cells' bounds. Having a particle always contribute completely to the one cell its midpoint is inside seems like an oversimplification. But it is acceptable since our density volume is coarse.

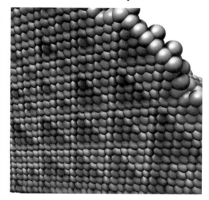

The simplifications this approach makes can result in visual artifacts in some special cases when the structure of the data set does not align with the density volume (see Figure 7.6). Then the volume data does not create a continuous representation of the particles. This relates to the implicit definition of the neighborhood evaluated for ambient occlusion factor given by the resolution of the density volume. Due to this simplified nature of occlusion data, long-range neighborhood information is not available.

Figure 7.6: Visible artifacts resulting from misaligned density volume data. Ambient occlusion factors are increased by ×3 to show the artifacts more clearly.

To remedy this issue we need to use a more sophisticated voxelization and volume evaluation strategy. More details are given by Staib et al. [2015]. The particles are voxelized into a higher resolution 3D texture, e.g., 256^3 or 512^3 voxels. When 3D textures are bound to a frame buffer, the geometry shader allows for each emitted point primitive to select a slice via the OpenGL variable `gl_Layer`. We employ the same particle data vertex buffer that is used for normal rendering. The geometry shader performs a slicing of each sphere. Given position and radius, point primitives are generated for each volume slice that intersects the sphere. We set the size of each primitive to match the maximum radius of the slice. Fragments that do not contribute to the slice are discarded. For all other fragments, the overlap between the sphere and the voxel cell is splatted to avoid aliasing effects. If the current voxel cell contains the complete sphere, then w is the fraction of space that the sphere occupies in the cell. If the voxel cell is completely contained in the sphere, then w is 1. The contributions of multiple spheres to a voxel cell are summed up using additive blending.

As a final step of the voxelization, a mipmap-pyramid is built for the voxel cone tracing by averaging. OpenGL provides the standard method `glGenerateMipmaps`. However, this method exhibits very poor performance. We believe the texture data is downloaded to the host, filtered, and re-uploaded to the GPU again. Utilizing the compute shader, even in a straightforward manner, the computation time can be reduced to 1/30 (cf. Listing 7.1).

Listing 7.1: GLSL compute shader performing 3D mipmap generation

```
1  #version 430
2  layout (local_size_x = 64, local_size_y=1, local_size_z=1) in;
3  uniform readonly layout(binding=0, RGBA8) image3D inputImage;
4  uniform writeonly layout(binding=1, RGBA8) image3D outputImage;
5  void main() {
6    ivec3 gid = ivec3(gl_GlobalInvocationID);
7    ivec3 pos = gid * 2;
```

```
 8   ivec3 posMax = pos + 1; // here we need to check that we stay
 9                           // inside the input image.
10   vec4 result;
11   for (int x = pos.x; x <= posMax.x; ++x)
12     for (int y = pos.y; y <= posMax.y; ++y)
13       for (int z = pos.z; z <= posMax.z; ++z)
14         result += imageLoad(inputImage, ivec3(x,y,z));
15   result /= 8.0;
16   imageStore(outputImage, gid, result);
17 }
```

The corresponding host code simply binds the two different mipmap levels of the same 3D texture and invokes the compute shader for all texel of the smaller level. Listing 7.2 shows the core function calls.

Listing 7.2: Core function calls invoking the GLSL compute for 3D mipmap generation

```
 1  while (res.x() >= 1.0f || res.y() >= 1.0f || res.z() >= 1.0f) {
 2    res *= 0.5f;
 3    glBindImageTexture(0, hVol,
 4      level, GL_TRUE, 0, GL_READ_ONLY, GL_RGBA8);
 5    glBindImageTexture(1, hVol,
 6      level + 1, GL_TRUE, 0, GL_WRITE_ONLY, GL_RGBA8);
 7    mipmapShader.Enable();
 8    mipmapShader.SetParameter("inputImage", 0);
 9    mipmapShader.SetParameter("outputImage", 1);
10    glDispatchCompute(ceil(res.x()/64), ceil(res.y()), ceil(res.z()));
11    mipmapShader.Disable();
12    glMemoryBarrier(GL_SHADER_IMAGE_ACCESS_BARRIER_BIT);
13    level++;
14  }
```

During glyph ray casting, again, the ambient occlusion term needs to be calculated. We perform voxel cone tracing in the previously voxelized scene in an iterative manner. Depending on a user selectable aperture angle, three cones, tightly aligned around the point's normal, are set up (cf. Figure 7.7, top).

For each cone, the sampling starts at a small offset from the glyphs surface point to avoid fetching values originating from the current glyph itself. While advancing in cone axis direction the mipmap level of the density volume is chosen corresponding to the cone diameter at that sampling position (cf. Figure 7.7, bottom). Voxel entries between two discrete mipmap levels are quadrilinearly interpolated by the GPU. Subsequently, we step one diameter step further, fetch a new sample, and accumulate the overall particle density in this direction. This process is repeated until a predefined cone length is reached. The length is typically set to 25% of the scene extend, but can, of course, be adjusted.

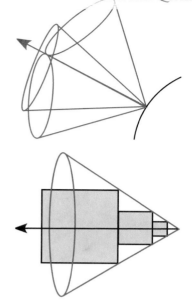

Figure 7.7: Top: setup of voxel cone tracing using three cones per AO factor evaluation. Bottom: sketch of the fundamental idea to choose coarser mipmap levels while moving along the cones axis.

The overall rendering cost obviously increases, since volume resolution increases, texture lookups increase drastically, and there are more fragment shader operations. Nevertheless, the image quality increases as well and sampling artifacts like those shown in Figure 7.6 no longer occur. On modern graphics cards the approach is still highly interactive.

The more general particle density volume representation and evaluation enables for even more sophisticated rendering modes. As one example, presented in Staib et al. [2015], transparency can be incorporated in this illumination model. Figure 7.8 shows a data set with transparency and ambient occlusion rendered interactively. The white spheres are modeled opaque while the green and blue spheres, depicting atoms forming linear crystal lattice defects, are rendered semi-transparent.

Volume rendering techniques integrate nicely with the voxel-based data set representation already in place. Instead of using a single-channel 3D volume storing particle density only, we splat colored contributions into a 3D RGBA texture. For this, we need to represent our particles in a corresponding manner.

The straightforward approach is to define a spherical region of constant volume in combination with simple transfer functions. The volume density is intuitively parametrized based on the opacity value of the particles' colors.

However, representing particles as spherical volumes with constant density results in a fuzzy border of the glyphs, as the length of the light ray inside the volume decreases toward zero. This counters our efforts to preserve the shape impression of the sphere. We introduce two additions to our density model to emphasize the spheres border. First, we do not use a spherical volume with constant density, but compute the density by the L_2^2 norm to the glyph's center. This strongly de-emphasizes the center of the glyph and the parameterization based on the particles opacity further improves the visibility of the actual border. As second addition, we explicitly remove a smaller concentric spherical sub-volume, resulting in a hollow sphere. Both additions are easily implementable in the volume evaluation of the sphere glyph.

During ray casting the glyph's surface base color, including its opacity value, is determined by the analytic solution of the volume rendering integral of this density model. The contribution of the particles to the RGBA density volume is now simply controlled by the particles' density. The alpha channel, directly representing the density, controls the light energy absorption during volume rendering integration. The color channels attenuate the non-absorbed light energy,

Figure 7.8:   Interactive rendering of semi-transparent spheres. Blending colored contributions to the ambient occlusion factors results in colored light (cf. spheres on the lower plane) [Staib et al., 2015].

effectively tinting the passing light. During the voxel cone tracing described above, the density accumulation needs to be replaced by volume-rendering-motivated alpha blending operation.

Obviously, now rendering semi-transparent glyphs, the particles need to be sorted in order to be rendered and blended correctly. We utilize an index buffer to sort the particle data depending on the current view. In each frame, the indices are updated by depth sorting if the viewpoint changes. A radix sort prefix-sum algorithm [Harris et al., 2007] can be used for this task. This algorithm is especially well suited as it is parallel in nature and has a linear run time, independent of the data order. The implementation can be trivially ported to any GPGPU language, including OpenCL, CUDA, and compute shaders.

Further techniques to increase the visual quality of the image can be used in combination with the presented techniques. Silhouettes of glyphs or glyph groups can be easily emphasized by a corresponding screen-space pass [Tarini et al., 2006]. Depth-cues like depth-of-field help us understand the 3D structure of particle data [Falk et al., 2013b]. From a visualization point of view, the beauty of images is not relevant, and so most works aim at removing or reducing visual artifacts. But an increase in visual quality helps us to gain better understanding of the data, especially in combination with abstracting and summarizing visual metaphors.

CHAPTER 8

Application-driven Abstractions

In the previous chapters, direct depictions of particle data sets were discussed. In this chapter, some examples for application-driven abstractions are given and discussed. The general idea is to create visualizations that show a special feature or property of the data. This can make the analysis of the data much easier. Often, the abstract visualization is based on a feature extraction, i.e., additional information is derived from the raw particle positions and other directly available properties. Subsequently, this information about secondary properties is visualized which might otherwise not be visible when looking at the particles. Simple properties that can be computed for all kinds of particle data sets comprise for example the spatio-temporal distribution of the particles or particle concentrations. However, particle data can occur in a lot of different application domains, and the application domain is often important since it gives semantics to the particles. This meaning of individual particles as well as their interactions has to be considered during the analysis as well as for the visualization.

One well-known example for such a derived particle representation instead of all individual particles is the depiction of deoxyribonucleic acid (DNA) as a ladder-like double helix. The nucleobase pairs form the rungs of the ladder while the rails of the ladder follow the sugar-phosphate backbones. This results in a very simple yet effective visualization of the geometric shape of the DNA. By showing only the overall structure instead of individual atoms, the visual clutter is greatly reduced, which makes it easier for the analyst to focus on the important properties. Another example would be the visualization of a particle-based fluid simulation. Instead of showing all particles that constitute the fluid (and which are necessary for the simulation), the analyst is often only interested in the fluid surface. Therefore, extracting and rendering a smooth surface based on the particles can be beneficial. If the analyst wants to study the internal dynamics of the fluid, techniques from flow visualization like streamlines can be used in order to create a more meaningful depiction that again reduces the clutter and shows interesting properties that might be hard to discern when looking at the raw particles.

Most of the examples in the rest of this chapter are from the domain of biomolecular visualization. The need for interactive, application-specific visualization of biomolecular data was identified early on [Langridge et al., 1981] and has been an active area of research ever since (see, e.g., Kozlíková et al. [2015], Krone et al. [2016]). In the following, the necessary basics in biochemistry and structural biology will be introduced briefly. Besides DNA, proteins are among the most widely studied biomolecules. They play an important role in all processes that occur in living organisms. Proteins are macromolecules that consist of smaller building blocks, namely amino acids. All amino acids share a common backbone part consisting of a carbon molecule (the so-called α-carbon), the amino group (–NH), and the carboxyl group (–COOH). The carboxyl group of an amino acid binds to the amino group of another amino acid, thereby creating a linear amino acid chain that forms the protein. Besides the backbone part, each amino acid has an individual side-chain that determines the physico-chemical properties of the amino acid. The side chains consist of only few atoms. The sequence of amino acids that forms the protein is called the *primary structure* of the protein. The chain will fold into an energetically favorable conformation that is stabilized by hydrogen bonds. This so-called *secondary structure* consists of structural motifs, namely helices, turns, and sheets. Unstructured parts of the chain are called random coils. The spatial arrangement of the secondary structure elements (helices, sheets, etc.) is called the *tertiary structure*. If two or more amino acid chains form a larger functional complex, it is called *quarternary structure*. For more details on protein folding, please refer to Richardson [1981].

From a simplified point of view, two main features are important for the function of a protein: the interface of the protein, which is defined by the amino acids that are at the surface (i.e., which are exposed to the environment), and the shape of the backbone, which depends on the folding. Consequently, these two features are also of interest for visualization in order to facilitate the analysis. In the following, visualization methods for these two features will be described as examples for application-driven abstractions.

8.1   SPLINE REPRESENTATIONS

Generally speaking, splines can form a smooth curve that follows a set of control points. Based on the type of spline, several properties can be obtained. For example, the continuity of the derivative depends on the order of the base polygon. A smooth spline in $\mathbb{R}^3$ requires at least cubic polygons. Cubic spline segments require a control polygon with four corners, i.e., control points. Another important property of the spline curve is whether it only approximates the control points or whether it passes through them. An example for a spline that passes through the control points is the cubic Hermite spline or the Catmull-Rom spline, which exhibits affine invariance and is thus more flexible in practice. An example for an approximating spline would be the cubic B-spline:

$$s(u) = d_0 N_0^3(u) + d_1 N_1^3(u) + d_2 N_2^3(u) + d_3 N_3^3(u) \tag{8.1}$$

with a knot vector T for four control points

$$T = u_0, u_1, \ldots, u_7 \tag{8.2}$$

and normalized B-Spline functions

$$N_i^0 = \begin{cases} 1 & \text{if } u_i < u \le u_i + 1 \\ 0 & \text{otherwise} \end{cases} \qquad (8.3)$$

$$\text{and} \quad N_i^r(u) = \frac{u - u_i}{u_{i+r} - u_i} N_i^{r-1}(u) + \frac{u_{i+1+r} - u}{u_{i+1+r} - u_{i+1}} N_{i+1}^{r-1}(u) \quad \text{for } 1 \le r < n . \qquad (8.4)$$

This spline consists of only a single segment between u_3 and u_4, but enlarging the knot vector as well as the *de Boor points* d will generate additional segments and, thus, an arbitrarily long curve. The curve evaluation basically slides a window over the relevant d_i for each segment, the correspondence being dictated by the knot vector: evaluating u requires searching an $l : u_l < u < u_{l+1}$ and then using $d_{l-3} \ldots d_l$. When advancing from one segment to the next, d_{l-3} is dropped, and d_{l+1} is added to the window. In this particular case, this ensures continuity even for the curve normals. A more detailed discussion of curves and surfaces can be found, e.g., in the textbook by Farin [2002].

Splines can be a useful representation for different phenomena in particle data sets. One of the most simple examples would be to show the path traveled by a particle in a dynamic data set. Usually, the path of each particle is stored using discreet time steps, i.e., the exact position of each particle at a certain point in time is stored. These discrete particle positions over time can be used as control points for the spline. Consequently, the resulting spline curve will show a smooth trajectory of the particle that approximates the actual path.

Another possibility for the utilization of a spline curve would be to use it to show the structure that is formed by a set of particles. An example for this would be the depiction of the structure of a linear macromolecule. Since the amino acid chain forming the protein is linear, it can be approximated by a spline which follows the backbone. The resulting representation will, hence, visualize the secondary and tertiary structure of the protein.

In order to obtain a spline following the backbone, the α-carbon atoms can be utilized as control points. These α-carbons are a good choice since they are located at the center of the backbone part of each amino acid. This will guarantee a spline curve that shows the path traced out by the backbone. The side chains are not shown by such a visualization. Consequently, the resulting abstract representation is reduced to the relevant parts and, therefore, induces much less visual clutter. In particular the side chains are not visible anymore. Furthermore, for the intended application—the analysis of the secondary structure—the exact atom positions and side chain orientations are typically not required.

In classical 3D computer graphics, a smooth spline curve usually has to be approximated by a set of straight line segments. The visual smoothness of the final, discretized curve depends on the number of line segments that are used to approximate each spline segment. On the other hand, the required computations also increase linearly with the number of line segments, which can lead to a decreased frame rate. Since the position of the α-carbon atoms can change for each frame, the discretization of the spline curve into line segments has to be interactive. In addition, uploading a lot of line segment data to the GPU can be a bottleneck for the interactive visualization as explained in Chapter 4. The goal is, therefore, to minimize the amount of data that has to be transferred to the GPU per frame while maximizing the number of line segments and maintaining interactive frame rates.

The computations for the line segments of the spline are independent of each other. That is, the line segment computation can be parallelized trivially and perfectly fits the SIMD architecture of the GPU. This also automatically minimizes the data that has to be transferred to the GPU since only the positions of the control points are needed. When using the GPU, there are two possibilities to compute the spline: in a shader that is part of the graphics pipeline (i.e., using OpenGL or Direct3D) or using an API that is independent of the graphics pipeline (e.g., CUDA or OpenCL). Since the problem of creating geometry for direct rendering maps well to the graphics pipeline, the first option is a sensible choice. New geometry can be created either using the *geometry shader* or the *tessellation shader*. While the geometry shader is more versatile, it is usually slower than the tessellation shader and has a lower limit of output vertices. The tessellation shader, on the other hand, is specifically designed for such tasks and is, therefore, the preferable option. In the following, a possible OpenGL implementation is described using the tessellation shader for generating a cubic spline curve from an arbitrary set of control points.

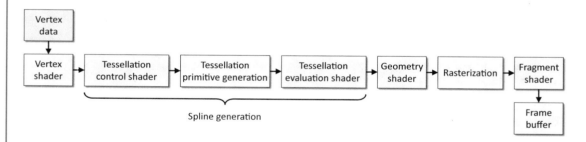

Figure 8.1: Simplified OpenGL pipeline. The tessellation state, which can be used for spline generation, consists of three stages: the tessellation control shader, the fixed-function primitive generation that generates the new primitives, and the tessellation evaluation shader.

Figure 8.1 shows a simplified version of the OpenGL pipeline. The programmable shader stages are colored in blue; fixed-function stages are colored in gray. The tessellation stage, which will be used to generate the splines in this example, is situated between the vertex shader and the geometry shader. It consists of three stages: the *tessellation control shader* (or *hull shader* in Direct X 11), the fixed-function *tessellation primitive generation* (or *tessellator* in DX) that generates the new primitives, and the *tessellation evaluation shader* (or *domain shader* in DX). The tessellation control shader defines the input vertices for the tessellation and the tessellation levels, i.e., the amount of new primitives that will be generated. The tessellation primitive generation creates the new primitives. The tessellation evaluation shader is similar to the vertex shader: it gets the tessellation coordinates of the current vertex and is responsible for the computation of their attributes. The tessellation evaluation shader also defines the type of primitive that will be tessellated and it can define the spacing between the newly created vertices.

The sample implementation will create a simple spline curve consisting of lines. Therefore, the *isolines* primitive will be used for the tessellation. First, all control points, i.e., the α-carbon positions, are uploaded to a shader storage buffer on the GPU. As mentioned above, the position vertices are uploaded only once to keep transfer times as low as possible. The number of spline segments should equal the number of control points, i.e., for *n* control points, the shader pipeline has to be invoked *n* times. Thus, *n* patches are rendered using glDrawArrays. Since the positions are already stored in the shader storage buffer, a geometry-less rendering can be used, meaning that no vertex positions are set during the draw call. The output of the tessellation control shader are the four control points defining the current (cubic) spline segment. That is, the four positions, corresponding to the α-carbon positions, are read from the shader storage buffer based on the invocation ID of the control shader and the primitive ID. The GLSL shader code of the control shader is shown in Listing 8.1. In this example, the tessellation level is set to 16, i.e., 16 line segments will be generated for each spline segment. For isolines, the inner tessellation level is not used. Note that each control point is reused by four consecutive spline segments in order to guarantee a continuous spline curve.

Listing 8.1: GLSL tessellation control shader for the spline generation

```
1   // define the number of vertices (input for tessellation evaluation)
2   layout( vertices=4) out;
3   // shader storage buffer with α-carbon positions (control points)
4   layout( packed, binding=2) buffer shader_data { vec4 positionsCa[]; };
5
6   void main() {
7       // write control point positions
8       // (4 output vertices, i.e., gl_InvocationID = [0..3]
9       gl_out[gl_InvocationID].gl_Position =
10          positionsCa[gl_PrimitiveID + gl_InvocationID];
11      // set outer tessellation level 0 to 1
12      // (one isoline is generated)
13      gl_TessLevelOuter[0] = float( 1);
14      // set outer tessellation level 1 to 16
15      // (16 line segments are generated per isoline)
16      gl_TessLevelOuter[1] = float( 16);
17  }
```

Afterward, the primitive generation will create the line segments of the isoline. The input of the tessellation evaluation shader are the control points defined in the control shader and the tessellation coordinate for the current vertex. Using this information, the correct position of the current vertex on the spline can be computed. Listing 8.2 shows the evaluation shader for a Catmull-Rom spline crossing all control points. An alternative would be to render a cubic B-spline, which only approximates the control points. The GLSL code for evaluating the B-spline equation can be found in Listing 8.3. This code takes advantage of a couple of optimizations: The spline is assumed to be uniform, i.e., the elements of the knot vector are given as $u_i = i$. Also, for uniform splines, the distances between knots are constant, so the normalized basis functions simplify to:

$$N_i^r(u) = \frac{u-i}{r} N_i^{r-1}(u) + \frac{i+1+r-u}{r} N_{i+1}^{r-1}(u) \quad \text{for} \quad 1 \leq r < n . \tag{8.5}$$

Since the local tessellation coordinate u will be in $[0, 1]$, an offset of size 3 will move it directly to the start of the defined segment $[u_3, u_4]$ (see above). Then the code evaluates the de Boor scheme by linearly interpolating between pairs of control points until a single point, the point on the curve, is left:

$$d_i^r(u) = (1 - \frac{u - u_{i+r}}{u_{i+n+1} - u_{i+r}}) d_i^{r-1} + \frac{u - u_{i+r}}{u_{i+n+1} - u_{i+r}} d_{i+1}^{r-1} \quad \text{for} \quad i = 3 - 3 = 0, \ldots, 3 - r .$$
$$\tag{8.6}$$

Since we know the interval starts at $i = 0$, we can directly resolve u_{i+r} to $r \ldots (3 - r + r = 3)$, while the distance $u_{i+n+1} - u_{i+r} = n + 1 - r$ is constant per recursion level and goes from 3 down to 1. This simplifies the implementation considerably.

Listing 8.2: GLSL tessellation evaluation shader for the spline generation. The sample code generates a Catmull-Rom spline.

```
1   // requested input from tessellator
2   layout( isolines , equal_spacing) in;
3
4   void main() {
5       // the 4 control points defined by the control shader
6       vec4 p0 = gl_in[0].gl_Position;
7       vec4 p1 = gl_in[1].gl_Position;
8       vec4 p2 = gl_in[2].gl_Position;
9       vec4 p3 = gl_in[3].gl_Position;
10
11      // the coordinate of the vertex within the current patch
12      float u = gl_TessCoord.x;
13      // equation for Catmull-Rom spline
14      gl_Position = 0.5 * ( (2.0 * p1) + (-p0 + p2) * u +
15          (2.0 * p0 - 5.0 * p1 + 4.0 * p2 - p3) * u * u +
16          (-p0 + 3.0 * p1- 3.0 * p2 + p3) * u * u * u);
17  }
```

Listing 8.3: GLSL tessellation evaluation shader snippet for the generation of a cubic B-spline. This snippet can be inserted in the tessellation evaluation shown in Listing 8.2 (replacing the Catmull-Rom equation).

```
1       // equation for a cubic B-spline
```

```
2      u += 3.0;
3      vec4 d10 = mix( p0, p1, ( u - 1.0) / 3.0);
4      vec4 d11 = mix( p1, p2, ( u - 2.0) / 3.0);
5      vec4 d12 = mix( p2, p3, ( u - 3.0) / 3.0);
6
7      vec4 d20 = mix( d10, d11, ( u - 2.0) / 2.0);
8      vec4 d21 = mix( d11, d12, ( u - 3.0) / 2.0);
9
10     gl_Position =  mix( d20, d21, ( u - 3.0));
```

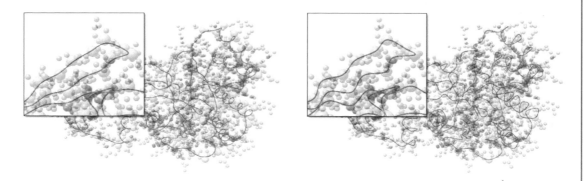

Figure 8.2: Comparison between the results of the two spline rendering methods for the backbone of a protein (PDB ID: 1VIS). Left: cubic B-spline, right: Catmull-Rom spline. The B-spline is much smoother, since it only approximates the positions of the α-carbons used as control points (shown as green spheres), whereas the Catmull-Rom spline crosses all control points. The rest of the atoms are rendered as gray spheres.

Figure 8.2 shows the resulting spline curve for a protein. In molecular visualization, this spline-based depiction of a protein is called *Cartoon* representation or *Ribbon* model (see, e.g., Carson [1987] or Richardson [1981]) and is commonly available in molecular visualization tools. The secondary structure elements are typically highlighted using additional geometry that follows the curve and decorates the spline. Helices and sheets are commonly depicted using a ribbon, while turns and random coils are usually rendered as tubes. Such geometry can also be generated on the GPU either by the tessellation shader using tringles as output primitive [Hermosilla et al., 2015] or by the geometry shader [Krone et al., 2008]. There is, however, no generally accepted mathematical definition of the exact appearance of the Cartoon model. Therefore, both cubic B-spines or Catmull-Rom splines can be used depending on the desired appearance. Example Cartoon renderings produced by the molecular visualization tool VMD [Humphrey et al., 1996] are depicted in Figure 8.3. While VMD does not use a GPU-accelerated implementation to generate the splines, the examples clearly show the different appearances of the two spline curves.

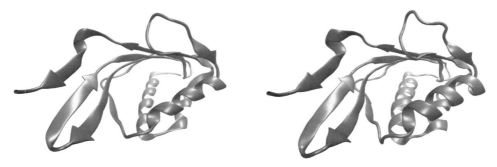

Figure 8.3: Cartoon renderings of a protein (PDB ID: 1OGZ) produced with VMD [Humphrey et al., 1996] using a cubic B-spline (left) and a Catmull-Rom spline (right).

As mentioned above, other examples where splines are a useful abstraction of particle data comprise the depiction of the DNA double helix or the visualization of particle trajectories (see, e.g., Ertl et al. [2014]). In both cases, the choice of the spline also depends on the requirements of the application and on the desired appearance and properties of the final depiction.

The current implementation uses a fixed number of line segments to approximate each spline segment. It is worth mentioning that the tessellation shader is typically also used to get an appropriate level of detail. The tessellation factor for the line segments, which is set in the control shader, can be defined based on the distance between the control points and the camera. Spline segments that are very close to the camera can be tessellated with a very high factor to get a smoother curve, whereas distant segments can be rendered with a lower tessellation factor. Since the distant segments will be much smaller, fewer line segments can be used without decreasing the image quality while increasing the rendering speed.

8.2   PARTICLE SURFACES

Extracting and visualizing a surface based on the particle positions can often give a good impression of a particle data set without showing all the individual particles. A natural example is a particle-based computational fluid dynamics simulation like smoothed particle hydrodynamics, where only the fluid surface is visualized (see, e.g., Yu and Turk [2013], Zhu and Bridson [2005]). An overview of surface reconstruction methods for fluid surfaces was given by Ihmsen et al. [2014]. Surface visualizations can also be used for other material boundaries or to delineate particle clusters (see, e.g., Müller et al. [2007], Scharnowski et al. [2013]).

In biomolecular visualization, surface representations are commonly used to show the surface of a molecule. There are different molecular surface definitions that show different properties of the visualized molecule. An extensive survey of molecular surface definitions and their computation was given by Kozlíková et al. [2015]. One popular example would be the *Solvent Excluded Surface* defined by Richards [1977], which depicts the boundary of a molecule with respect to a solvent molecule of a certain size. The solvent is approximated by a spherical probe. The idea is that points which are not reachable by the probe, i.e., the solvent, are enclosed by the surface.

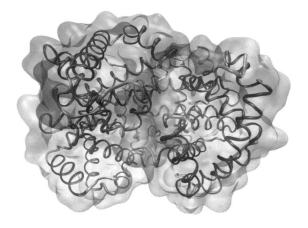

Figure 8.4: Visualization of a hemoglobin protein (PDB ID: 1HBB). The semi-transparent molecular surface and the spline, illustrating the secondary structure of the protein, are both derived only from the atomic positions.

Another popular surface definition for particle data was given by Blinn [1982]. Here, each particle is represented by a (typically radially symmetric) density function that depends on the distance to the center of the particle. For each point $r \in \mathbb{R}^3$, the density contribution of all particles is summed up. From this density field an implicit surface can be extracted. These implicit surfaces are also known as *blobby surfaces*, *Metaballs*, or *convolution surfaces*. Convolution surfaces provide smooth surfaces that are widely applicable to different application areas depending on the kernel function which is applied to each particle. Usually, a Gaussian function is used as density kernel function per particle. In this case, convolution surfaces are also called Gaussian (density) surfaces. Gaussian density surfaces can approximate the electron density distribution of a molecule. Furthermore, they can also approximate the Solvent Excluded Surface. An example is shown in Figure 8.4, where a semi-transparent Gaussian surface is used to visualize a protein in combination with a spline representation.

A fast GPU-accelerated implementation to compute and visualize the Gaussian surface of a protein was presented by Krone et al. [2012]. The general idea is to calculate the density values on a regular volumetric grid that encloses all atoms. The number of voxels—i.e., the size of one grid cell—determines the accuracy and image quality. However, using a very fine grid will naturally also lead to higher computational cost and will negatively affect the performance since the computational complexity will increase with a factor of $\mathcal{O}(n^3)$ with decreasing grid cell size. For biomolecular data, a side length of $1\,\text{Å}$ ($= 0.1\,\text{nm}$) for the grid cells is usually sufficient to provide sub-atomic detail as the smallest atom, hydrogen, has a van der Waals radius of $1.2\,\text{Å}$.

When using a Gaussian density kernel, the density ρ at an arbitrary point $\boldsymbol{v} \in \mathbb{R}^3$ can be calculated as

$$\rho(\boldsymbol{v}) = \sum_{i=1}^{n} e^{\frac{-|\boldsymbol{v}-\boldsymbol{v}_i|^2}{2\alpha^2}} , \qquad (8.7)$$

where n is the number of atoms, $\boldsymbol{v}_i$ is the position of atom i, and $\alpha = r_i \cdot \omega$ is the radius r_i of atom i radius multiplied with a weighting factor ω.

For m points in space—i.e., m voxels—summing up the density contribution of all n atoms will result in a runtime complexity of $\mathcal{O}(n \cdot m)$, which will quickly become too costly for an interactive computation. This computation, however, is necessary for dynamic data, where the atom can move for each frame. In consequence, the density volume has to be updated in real time during rendering. To reduce the computational complexity, the fact that a Gaussian function decays rapidly toward zero can be exploited. The density contribution of an atom that has a large distance to point $\boldsymbol{v}$ is negligible. As shown in Equation 8.7, the density contribution of an atom i to point $\boldsymbol{v}$ also depends on a constant weighting factor ω and the atom radius r_i. Therefore, a fixed cutoff radius r_c can be found based on the factors ω and the maximum atom radius $\max(r_i)$ of all atoms in the data set. Only atoms that are within the distance r_c to $\boldsymbol{v}$ are considered for the calculation of the density $\rho(\boldsymbol{v})$. That is, for each voxel position $\boldsymbol{v}$, the atoms within the fixed search radius r_c have to be found. This can be done efficiently using the fixed-radius nearest neighbor search algorithm explained in Chapter 6. Obviously, small cutoff radii will induce visible errors since close atoms might be omitted that would have a visible impact on the density ρ. A cutoff radius $2 \cdot \alpha \geq r_c \geq 4 \cdot \alpha$ will usually be sufficiently large (see Krone et al. [2012]).

Since the density values of all voxels are independent of each other, they can be computed in parallel without write conflicts or the need for synchronization between threads. This strategy is called a *gathering approach* since the density contributions of all neighboring atoms are gathered for each point $\boldsymbol{v}$. The alternative would be a *scattering approach* where the density contribution to each voxel within r_c is written for each atom i. While the latter approach would not require the neighbor search, it would potentially cause write conflicts since two atoms i and j could contribute to the same voxel position $\boldsymbol{v}$. Therefore, the first approach is usually preferable for a GPU-accelerated implementation. Listing 8.4 shows an excerpt of the CUDA source code where each thread computes the density of one voxel.

Listing 8.4: Source code excerpt from the CUDA kernel computing of the density at position coord, i.e., the voxel position $\boldsymbol{v}$

```
1   float3 coord = ...; // position v in R^3 of the current voxel
2
3   // --- loop over all neighboring cells; idx = cell index ---
4   // ...
5       for ( uint pIdx = cellStart[idx]; pIdx <= cellEnd[idx]; pIdx++) {
6           float4 particle  = particlePosition[pIdx];
7           float dx = coord.x - particle.x; // distance in x-direction
```

```
8          float dy = coord.y - particle.y; // distance in y-direction
9          float dz = coord.z - particle.z; // distance in z-direction
10         float dxyz2 = dx * dx + dy * dy + dz * dz;
11         densityval += exp2f(dxyz2 / particle.w); // w stores -2*a^2
12     }
13  // ...
```

In order to speed up the computation, the workload per thread can be optimized further as shown in Listing 8.5. The idea is to compute the density values of several adjacent voxels in the same thread. The voxels all have the same x- and y-coordinate in the density grid, only the z-coordinate is incremented. Consequently, the combined neighborhood for all voxels will usually be only slightly larger since the size of the grid cells (`densityGridSpacing`) is typically smaller than the radius of an atom. That way, not only the neighborhood search has to be executed only once for all voxels, the variable `dxy2` can also be re-used for all voxels. However, the number of voxels that will be computed in one thread has to be limited to a reasonable number such that the neighborhood of the voxels largely overlaps. Using eight voxels has been experimentally found to be a good number [Krone et al., 2012]. The optimal value, however, has to be carefully evaluated based on both problem size and hardware.

Listing 8.5: Source code excerpt of the optimized CUDA kernel for the computation of the density at position `coord`. By computing the densities of adjacent voxels in the same thread, neighbor information and computation results can be re-used.

```
1  float3 coord = ...; // position v in R^3 of the current voxel
2
3  // --- loop over all neighboring cells; idx = cell index ---
4  // --- note: use the combined neighborhood of all cells   ---
5  // ...
6      for ( uint pIdx = cellStart[idx]; pIdx <= cellEnd[idx]; pIdx++) {
7          float4 particle  = particlePosition[pIdx];
8          float dx = coord.x - particle.x; // distance in x-direction
9          float dy = coord.y - particle.y; // distance in y-direction
10         float dxy2 = dx * dx + dy * dy;
11         // first distance in z-direction
12         float dz_1 = coord.z - particle.z;
13         densityval_1 += exp2f((dxy2 + dz_1*dz_1) / particle.w);
14         // second distance in z-direction
15         float dz_2 = dz_1 + densityGridSpacing.z;
16         densityval_2 += exp2f((dxy2 + dz_2*dz_2) / particle.w);
17         ...
18     }
19  // ...
```

After computing the density volume, an appropriate isosurface can be extracted for rendering. The straightforward approach would be a direct visualization of the isosurface using fast GPU-based volume ray marching. This approach provides high image quality and correct transparency without explicit sorting. The performance of ray marching, however, highly depends on the image resolution but can be done in parallel for all rays. A CUDA implementation of a simple volume ray marching can be found in the samples of the CUDA Toolkit provided by NVIDIA. An alternative to direct volume rendering with ray marching would be to extract a tessellated isosurface using the Marching Cubes algorithm [Lorensen and Cline, 1987]. The triangle mesh extracted by Marching Cubes will still provide good image quality and has several advantages compared to ray casting. The surface mesh can be rendered efficiently from multiple viewpoints, e.g., for stereoscopic rendering. It can also be used for simple surface analyses. For example, the approximate surface area can be calculated by summing up the individual triangle areas. A surface mesh is also advantageous if permanent storage is required since it can be stored on disk more compactly than the volume data. Marching Cubes (and related algorithms like Marching Tetrahedra) can be trivially parallelized for fast computation on the GPU. The triangles within each cube of $2 \times 2 \times 2$ voxels can be extracted efficiently in parallel using a lookup table. Based on the eight voxel values, this lookup table yields the triangles within this cube. Only the exact vertex positions have to be computed using linear interpolation between two voxel positions. The samples of the CUDA Toolkit also include a parallel Marching Cubes implementation which will be outlined in the following.

As mentioned above, the triangles within one cube can be computed in parallel. The problem is that between zero and four triangles will be generated per cube. These triangles have to be stored prior to rendering. The naïve approach would be to reserve enough memory per cube to enable potentially writing all four triangles. This would result is a huge waste of memory since typically only 15% of the cubes actually contains parts of the isosurface of a molecular data set. In addition, the scattered data would make an efficient rendering difficult. A memory-efficient approach would be to use dynamic data structures; however, while this is possible using CUDA, it is typically quite slow. Therefore, a different strategy is preferable. First, all cubes are classified in parallel using the abovementioned lookup table to obtain the number of triangles that will have to be generated for this cube (see, e.g, http://paulbourke.net/geometry/polygonise/). The number of triangles per cube is written to an array. Using a prefix sum, the total number of triangles can be computed (e.g., using the `thrust::scan_exclusive` function provided by the Thrust library). For a cube i, this also gives the number of triangles that will be generated for all previous cubes $1 \ldots i - 1$. That is, the prefix sum allows us to compute a compacted index for each triangle vertex. After allocating the required memory for all triangles, another kernel is executed that classifies the cubes again and writes the actual vertex positions for the triangle mesh to the output array using the previously computed indices. For a high-quality shading, per-vertex normals have to be computed. They can either be derived from the volume using central differences or as the average of the per-vertex normals of the triangles connected to the vertex.

Similar to the particle densities, additional information can be stored in the density volume, e.g., the index of the closest particle or the average color of all particles that contribute to this voxel (weighted by their respective density contribution). This information can be used to compute

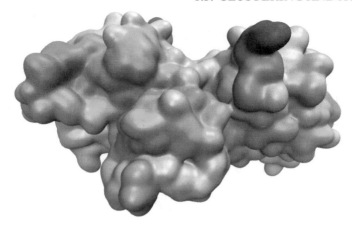

Figure 8.5: Molecular surface of a protein (PDB ID: 3PPJ) colored by the temperature factor. The CUDA-accelerated Gaussian molecular surface definition is available in MegaMol [Grottel et al., 2015] and VMD [Humphrey et al., 1996] as *QuickSurf* representation [Krone et al., 2012].

attributes like per-vertex colors or atom contacts for the surface. Figure 8.5 shows an example for a colored Gaussian surface computed for a protein.

Using volumetric representations can quickly lead to a very high memory consumption for large data if a small grid cell size is used. This can lead to volume sizes that exceed the available GPU memory. In this case, the Gaussian surface computations can be trivially divided into subblocks. That is, subvolumes of the whole density volume can be computed. The only caveat is that the subvolumes have to have an overlap of one voxel. Otherwise, the extracted isosurface will exhibit gaps.

8.3   CLUSTERING AND AGGREGATION

Clustering and aggregation can be viable tools for the visual analysis of particle data. In this section, several possible application-specific visualizations for biomolecular data sets will be discussed briefly without going into technical details. Clustering can help us to find similarities by grouping particles according to some property, it can reduce visual clutter by only rendering a proxy geometry for larger clusters of particles, or it can be used to show only prominent features of interest by filtering clusters according to a certain property. Aggregation can for example show the overall probability of presence if the particle positions are temporally aggregated or it can provide averaged properties.

A concrete example for clustering would be the extraction of main paths of water molecules [Bidmon et al., 2008]. Here, the path of each water molecule which crosses a certain region of interest (ROI) is tracked for the entire simulation. These trajectories can then be

Figure 8.6: Clustering of solvent path lines described by Bidmon et al. [2008]. The trajectories of all water molecules that cross a certain region of interest are drawn, which results in high visual clutter (left). The main paths are found by clustering similar paths, resulting in a much clearer depiction (right). The width of the tubes represents the size of the cluster, i.e., the number of clustered trajectories. From [Bidmon et al., 2008].

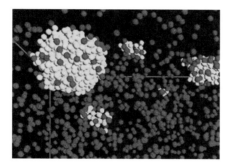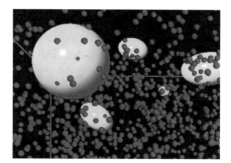

Figure 8.7: Clustering of particles according to spatial density. The clusters (cyan) are rendered as ellipsoids which enclose all particles that belong to this cluster. From [Grottel et al., 2007].

clustered according to the particle movement near the ROI. That is, clusters of molecules that follow a similar path found. Additional properties like the velocity of the particle can be considered as additional criteria for the clustering. The resulting main paths formed by the largest clusters can for example be visualized using the GPU-accelerated spline rendering described in Section 8.1. This not only shows the information about the main paths obtained by the clustering, it also highly reduces visual clutter since only one spline is drawn for all members of the cluster instead of all the individual trajectories. Figure 8.6 shows a comparison between visualizing all paths and depicting only the main paths found by clustering.

Clustering can also be very useful without temporal information. A clustering based on particle densities can for example be used to create a level-of-detail hierarchy for fast render-

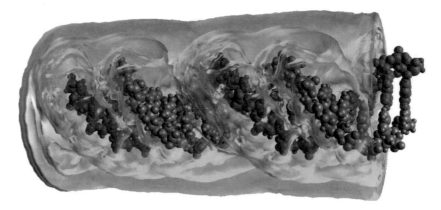

Figure 8.8: Layers of different ion densities around DNA. The density volume containing the ion densities was obtained by spatio-temporal aggregation. From [Ertl et al., 2014].

ing [Le Muzic et al., 2014], as explained in Section 5.2. A possibility would be to use established algorithms like DBSCAN [Ester et al., 1996] or k-means [MacQueen, 1967] to find agglomerates of particles. Since clustering often requires finding neighboring particles, it can be accelerated using the fast GPU-based fixed-radius neighbor search explained in Chapter 6. For visualization, the clusters can be replaced by a simple proxy like a sphere or an ellipsoid approximating the shape of the cluster. This will be less costly to render than all individual spheres [Grottel et al., 2007] (see Figure 8.7). When rendering macromolecular data, clusters can also be defined based on the inherent structure defined by the atoms. For example, all atoms forming one amino acid can be clustered and rendered as a sphere (see, e.g., Krone et al. [2009], Le Muzic et al. [2014]).

Temporal aggregation has also been successfully applied to biomolecular data. An example is the spatio-temporal aggregation of ion positions and velocities around a DNA double helix by Ertl et al. [2014]. The result is a density field that stores the probability of presence for the ions as well as the average ion velocity at a certain position. For each time step, the atom positions and velocities can be written to a volumetric grid using the CUDA-accelerated algorithm explained in Section 8.2. Subsequently, all density volumes are summed up and the voxel values are normalized by dividing them by the number of time steps. The resulting aggregated density grid can be visualized using GPU-accelerated rendering methods like volume ray marching (see Figure 8.8) or streamline extraction. This allows us to get detailed insights into the average properties of the ions.

CHAPTER 9

Summary and Outlook

This lecture has shown practical approaches for the rendering of particle data. The introduction described application areas and data and gave a brief historical overview of evolving hardware and the corresponding improvements in technique. First, the basic hybrid image/object-order GPU-based glyph ray casting was detailed, also showing approaches for optimizing data handling as well as improving visual quality for easier interpretation of data. Next, the acceleration of the image-order approach for whole scenes via a uniform grid was shown together with a hierarchical LoD scheme for particles. The hierarchy was also employed to reduce memory footprint by quantization. These approaches are complemented by application-specific abstractions. They simplify data and its interpretation and can highlight properties specific to some analysis questions, like the internal structure of a protein as well as its interface to the environment. The requirement of real-time interactive exploration was honored throughout all topics; scaling was ensured through abstraction and thus simplification of the represented data. All employed visual metaphors correspond to textbook knowledge (i.e., to representations established in the application domain), thus closing the circle to potential users.

The presented approaches are, of course, only a starting point for further improvement. With increasing computational power, the available data sets will steadily grow in size and complexity, so scalability will remain an important factor. In turn the impracticability of serialization will become more widespread: storage space cannot increase at the same pace as the generation of simulation results. Furthermore, serialization is a performance issue as it stalls the simulation, often requiring also the gathering of all data distributed on a cluster on a much smaller storage backend. Therefore, the importance of *in situ* visualization will further increase. *In situ* requires us to make a choice of where to split the visualization pipeline and insert serialization. The rendering output can be produced in a distributed manner and the result captured, but the reduced interactivity is often not acceptable. Intermediate solutions like *volumetric depth images* exist, but in the long term an *in situ* preprocessing/abstraction of data for rendering it interactively later will be a more flexible solution. However, domain-specific knowledge will be required to optimally reduce the available data. Finding an alternative solution that does not bias the output would be preferable, but how such a generic method would look is unclear at best. A topic that was not covered by our book is visualization using multiple GPUs. At least in theory, most algorithms discussed in this book can be implemented rather straightforwardly to run in parallel on multiple GPUs. One caveat is that such implementations usually require us to composit the partial images of all

GPUs into one. While this is also quite straightforward and computationally cheap for opaque objects, it can be more challenging for semi-transparent objects.

Last, the exploration itself is a scalability problem. With increasing size of the data, the user will be lost when first faced with a new data set. It might even be unclear in which part of the data an interesting phenomenon might manifest itself, so it must be made easier for the user to decide which parts of the data he/she might focus on. We can deduce a very difficult, additional requirement from this: visualizations need abstractions that are consistent across scales. Neither the base-line representation nor its abstractions must generate false features, because they will cost time to explore and verify and ultimately discard. Conversely, minute details should be preserved all the way up to the highest abstractions. It is here where we expect the biggest potential of improvement over the techniques that are in use today. We expect that such abstractions will require new algorithms that analyze the data in order to extract features of interest on multiple scales. These features could then be used to automatically generate abstract visualizations that are tailored to the application. We also envision that the visual abstractions can be influenced by the user choosing the degree of detail as well as steering the extraction of interesting features. This would result in a large step toward a user-driven interactive visual analysis of large particle data.

Bibliography

T. Akenine-Möller. Fast 3D triangle-box overlap testing. *Journal of Graphics Tools*, 6(1), pages 29–33, 2001. DOI: 10.1080/10867651.2001.10487535. 58

J. Amanatides and A. Woo. A fast voxel traversal algorithm for ray tracing. In *EuroGraphics 1987*, pages 3–10, 1987. http://citeseerx.ist.psu.edu/viewdoc/summary?doi=10.1.1.42.3443. 49, 55

C. Bajaj, P. Djeu, V. Siddavanahalli, and A. Thane. TexMol: Interactive visual exploration of large flexible multi-component molecular complexes. In *Proc. of the Conference on Visualization '04*, pages 243–250, 2004. DOI: 10.1109/visual.2004.103. 7, 8

J. L. Bentley. Multidimensional binary search trees used for associative searching. *Communications of the ACM*, 18(9), pages 509–517, 1975. ISSN 0001-0782. http://doi.acm.org/10.1145/361002.361007. DOI: 10.1145/361002.361007. 66

H. M. Berman, J. Westbrook, Z. Feng, G. Gilliland, T. N. Bhat, H. Weissig, I. N. Shindyalov, and P. E. Bourne. The protein data bank. *Nucleic Acids Research*, 28(1), pages 235–242, 2000. ISSN 0305-1048, 1362-4962. http://www.pdb.org. DOI: 10.1093/nar/28.1.235. 51

K. Bidmon, S. Grottel, F. Bös, J. Pleiss, and T. Ertl. Visual abstractions of solvent pathlines near protein cavities. *Computer Graphics Forum*, 27(3), pages 935–942, 2008. DOI: 10.1111/j.1467-8659.2008.01227.x. 95, 96

J. F. Blinn. Models of light reflection for computer synthesized pictures. *SIGGRAPH Comput. Graph.*, 11(2), pages 192–198, July 1977. ISSN 0097-8930. http://doi.acm.org/10.1145/965141.563893. DOI: 10.1145/965141.563893. 11

J. F. Blinn. A generalization of algebraic surface drawing. *ACM Transactions on Graphics*, 1(3), pages 235–256, 1982. ISSN 0730-0301. DOI: 10.1145/357306.357310. 91

M. Botsch and L. Kobbelt. High-quality point-based rendering on modern GPUs. In *Pacific Conference on Computer Graphics and Applications*, pages 335–343. IEEE, 2003. DOI: 10.1109/pccga.2003.1238275. 5

M. Carson. Ribbon models of macromolecules. *Journal of Molecular Graphics*, 5(2), pages 103–106, 1987. ISSN 0263-7855. http://www.sciencedirect.com/science/article/pii/0263785587800103. DOI: 10.1016/0263-7855(87)80010-3. 89

R. O. Dror, R. M. Dirks, J. Grossman, H. Xu, and D. E. Shaw. Biomolecular simulation: A computational microscope for molecular biology. *Annual Review of Biophysics*, 41(1), pages 429–452, 2012. http://dx.doi.org/10.1146/annurev-biophys-042910-155245. DOI: 10.1146/annurev-biophys-042910-155245. 2

W. Eckhardt, A. Heinecke, R. Bader, M. Brehm, N. Hammer, H. Huber, H.-G. Kleinhenz, J. Vrabec, H. Hasse, M. Horsch, M. Bernreuther, C. Glass, C. Niethammer, A. Bode, and H.-J. Bungartz. 591 tflops multi-trillion particles simulation on supermuc. In *International Supercomputing Conference (ISC) Proceedings 2013*, volume 7905 of *Lecture Notes in Computer Science*, pages 1–12, Heidelberg, Germany, June 2013. Springer. DOI: 10.1007/978-3-642-38750-0_1. 1

T. Ertl, M. Krone, S. Kesselheim, K. Scharnowski, G. Reina, and C. Holm. Visual analysis for space-time aggregation of biomolecular simulations. *Faraday Discussions*, 169, pages 167–178, 2014. ISSN 1364-5498. DOI: 10.1039/c3fd00156c. 90, 97

M. Ester, H. P. Kriegel, J. Sander, and X. Xu. A density-based algorithm for discovering clusters in large spatial databases with noise. In *Internatonal Conference on Knowledge Discovery and Data Mining (KDD'96)*, pages 226–231, 1996. 97

M. Falk, M. Krone, and T. Ertl. Atomistic visualization of mesoscopic whole-cell simulations using ray-casted instancing. *Computer Graphics Forum*, 32(8), pages 195–206, 2013a. http://dx.doi.org/10.1111/cgf.12197. DOI: 10.1111/cgf.12197. 74

M. Falk, M. Krone, and T. Ertl. Atomistic visualization of mesoscopic whole-cell simulations using ray-casted instancing. *Computer Graphics Forum*, 32(8), pages 195–206, 2013b. ISSN 1467-8659. DOI: 10.1111/cgf.12197. 49, 50, 51, 58, 81, 109

G. Farin. *Curves and Surfaces for CAGD: A Practical Guide*. Morgan Kaufmann Publishers Inc., San Francisco, CA, 5th ed., 2002. ISBN 1-55860-737-4. 85

B. E. Feldman, J. F. O'brien, and O. Arikan. Animating suspended particle explosions. *ACM Transactions on Graphics (TOG)*, 22(3), pages 708–715, 2003. DOI: 10.1145/882262.882336. 5

R. Fraedrich, J. Schneider, and R. Westermann. Exploring the millennium run—scalable rendering of large-scale cosmological datasets. *IEEE Transactions on Visualization and Computer Graphics*, 15(6), pages 1251–1258, 2009. DOI: 10.1109/tvcg.2009.142. 6

D. Frenkel and B. Smit. *Understanding Molecular Simulation: From Algorithms to Applications*. Academic Press, 2nd ed., 2001. ISBN 978-0-12-267351-1. DOI: 10.1063/1.881812. 65

F. Gieseke, J. Heinermann, C. Oancea, and C. Igel. Buffer k-d trees: Processing massive nearest neighbor queries on gpus. In *Proc. of the 31st International Conference on Machine Learning (ICML)*, pages 172–180, 2014. 67

R. Gingold and J. Monaghan. Smoothed particle hydrodynamics: Theory and application to non-spherical stars. *Monthly Notices Royal Astronomical Society*, 181, pages 375–389, 1977. DOI: 10.1093/mnras/181.3.375. 65

P. Goswami, F. Erol, R. Mukhi, R. Pajarola, and E. Gobbetti. An efficient multi-resolution framework for high quality interactive rendering of massive point clouds using multi-way kd-trees. *The Visual Computer*, 29(1), pages 69–83, 2013. ISSN 1432-2315. DOI: 10.1007/s00371-012-0675-2. 5

S. Green. Particle simulation using CUDA. Technical report, NVIDIA Corp., 2012. 67

M. Gross and H. Pfister, editors. *Point-Based Graphics*. Morgan Kaufmann Publishers, 2007. DOI: 10.1016/b978-0-12-370604-1.x5000-7. 5

S. Grottel. *Point-based Visualization of Molecular Dynamics Data Sets*. Ph.D. thesis, Visualisierungsinstitut der Universität Stuttgart, 2012. 17, 21

S. Grottel, G. Reina, J. Vrabec, and T. Ertl. Visual verification and analysis of cluster detection for molecular dynamics. *IEEE Transactions on Visualization and Computer Graphics*, 13(6), pages 1624–1631, 2007. ISSN 1077-2626. DOI: 10.1109/tvcg.2007.70614. 96, 97

S. Grottel, G. Reina, and T. Ertl. Optimized data transfer for time-dependent, GPU-based glyphs. In *IEEE Pacific Visualization Symposium (PacificVis 2009)*, pages 65–72, 2009a. DOI: 10.1109/pacificvis.2009.4906839. 15, 31, 34

S. Grottel, G. Reina, and T. Ertl. Optimized data transfer for time-dependent, gpu-based glyphs. In *Proc. of IEEE Pacific Visualization Symposium 2009*, pages 65–72, 2009b. 10.1109/PACIFICVIS.2009.4906839. DOI: 10.1109/pacificvis.2009.4906839. 25

S. Grottel, G. Reina, C. Dachsbacher, and T. Ertl. Coherent culling and shading for large molecular dynamics visualization. *Computer Graphics Forum*, 29(3), pages 953–962, 2010a. DOI: 10.1111/j.1467-8659.2009.01698.x. 42, 47, 72, 74

S. Grottel, G. Reina, T. Zauner, R. Hilfer, and T. Ertl. Particle-based rendering for porous media. In *Annual SIGRAD Conference*, pages 45–51, 2010b. 4, 16

S. Grottel, P. Beck, C. Müller, G. Reina, J. Roth, H.-R. Trebin, and T. Ertl. Visualization of electrostatic dipoles in molecular dynamics of metal oxides. *IEEE Transactions on Visualization and Computer Graphics*, 18(12), pages 2061–2068, 2012a. DOI: 10.1109/tvcg.2012.282. 2

S. Grottel, M. Krone, K. Scharnowski, and T. Ertl. Object-space ambient occlusion for molecular dynamics. In *IEEE Pacific Visualization Symposium (PacificVis 2011)*, pages 209–216. IEEE, 2012b. DOI: 10.1109/pacificvis.2012.6183593. 74, 75

S. Grottel, M. Krone, C. Müller, G. Reina, and T. Ertl. MegaMol—A prototyping framework for particle-based visualization. *IEEE Transactions on Visualization and Computer Graphics*, 21(2), pages 201–214, 2015. ISSN 1077-2626. http://www.megamol.org. DOI: 10.1109/tvcg.2014.2350479. 95, 109

S. Gumhold. Splatting illuminated ellipsoids with depth correction. In *Vision, Modeling, and Visualization*, pages 245–252, 2003. 9, 11

M. Harris, S. Sengupta, and J. D. Owens. Parallel prefix sum (scan) with CUDA. In H. Nguyen, ed., *GPU Gems 3*, chapter 39, pages 851–876. Addison Wesley, August 2007. 81

P. Hermosilla, V. Guallar, A. Vinacua, and P.-P. Vázquez. Instant Visualization of Secondary Structures of Molecular Models. In *Eurographics Workshop on Visual Computing for Biology and Medicine*. The Eurographics Association, 2015. ISBN 978-3-905674-82-8. DOI: 10.2312/vcbm.20151208. 89

R. Hoetzlein. Fast fixed-radius nearest neighbor search on the GPU: Interactive million-particle fluids. In *Game Developer Conference*. NVIDIA, 2014. 70

M. Hopf and T. Ertl. Hierarchical splatting of scattered data. In *IEEE Visualization 2003*, 2003. DOI: 10.1109/mcg.2004.7. 6, 49, 59, 61, 62, 63, 64

D. R. Horn, J. Sugerman, M. Houston, and P. Hanrahan. Interactive k-d tree gpu raytracing. In *Proc. of the 2007 Symposium on Interactive 3D Graphics and Games*, I3D '07, pages 167–174, 2007. DOI: 10.1145/1230100.1230129. 67

W. Humphrey, A. Dalke, and K. Schulten. VMD—Visual molecular dynamics. *Journal of Molecular Graphics*, 14, pages 33–38, 1996. DOI: 10.1016/0263-7855(96)00018-5. 7, 89, 90, 95

M. Ihmsen, J. Orthmann, B. Solenthaler, A. Kolb, and M. Teschner. SPH Fluids in Computer Graphics. In *Eurographics 2014—State of the Art Reports*. The Eurographics Association, 2014. DOI: 10.2312/egst.20141034. 90

T. Klein and T. Ertl. Illustrating magnetic field lines using a discrete particle model. In *Vision, Modeling, and Visualization*, pages 387–394, 2004. 9, 11

A. Knoll, Y. Hijazi, A. Kensler, M. Schott, C. Hansen, and H. Hagen. Fast ray tracing of arbitrary implicit surfaces with interval and affine arithmetic. *Computer Graphics Forum*, 28(1), pages 26–40, 2009. DOI: 10.1111/j.1467-8659.2008.01189.x. 9

B. Kozlíková, M. Krone, N. Lindow, M. Falk, M. Baaden, D. Baum, I. Viola, J. Parulek, and H.-C. Hege. Visualization of biomolecular structures: State of the art. In *Eurographics Conference on Visualization—STARs*, pages 61–81, 2015. DOI: 10.2312/eurovisstar.20151112. 83, 90

M. Krone, K. Bidmon, and T. Ertl.   GPU-based visualisation of protein secondary structure. In *Theory and Practice of Computer Graphics*, volume 8, pages 115–122, 2008. DOI:
10.2312/LocalChapterEvents/TPCG/TPCG08/115-122. 89

M. Krone, K. Bidmon, and T. Ertl. Interactive visualization of molecular surface dynamics. *IEEE
Transactions on Visualization and Computer Graphics*, 15(6), pages 1391–1398, 2009. ISSN
1077-2626. DOI: 10.1109/tvcg.2009.157. 97

M. Krone, J. E. Stone, T. Ertl, and K. Schulten.   Fast visualization of Gaussian density surfaces for molecular dynamics and particle system trajectories.   In
EuroVis—Short Papers, volume 1, pages 67–71, 2012.   ISBN 978-3-905673-91-3. DOI:
10.2312/PE/EuroVisShort/EuroVisShort2012/067-071. 91, 92, 93, 95

M. Krone, B. Kozlíková, N. Lindow, M. Baaden, D. Baum, J. Parulek, H.-C. Hege, and I. Viola.
Visual analysis of biomolecular cavities: State of the art. *Computer Graphics Forum*, 35(3),
pages 527–551, 2016. ISSN 1467-8659. DOI: 10.1111/cgf.12928. 83

A. Lagae and P. Dutré. Compact, fast and robust grids for ray tracing. *Computer Graphics Forum*,
27(4), pages 1235–1244, 2008. DOI: 10.1111/j.1467-8659.2008.01262.x. 52

O. D. Lampe, I. Viola, N. Reuter, and H. Hauser. Two-level approach to efficient visualization of protein dynamics. *IEEE Transactions on Visualization and Computer Graphics*, 13(6),
pages 1616–1623, 2007. DOI: 10.1109/tvcg.2007.70517. 61

R. Langridge, T. E. Ferrin, I. D. Kuntz, and M. L. Connolly. Real-time color graphics in stud-
ies of molecular interactions. *Science*, 211(4483), pages 661–666, Feb. 1981. ISSN 0036-
8075, 1095-9203. http://science.sciencemag.org/content/211/4483/661. DOI:
10.1126/science.7455704. 83

M. Le Muzic, J. Parulek, A. Stavrum, and I. Viola. Illustrative visualization of molecular re-
actions using omniscient intelligence and passive agents. *Computer Graphics Forum*, 33(3),
pages 141–150, 2014. ISSN 1467-8659. http://dx.doi.org/10.1111/cgf.12370. DOI:
10.1111/cgf.12370. 49, 59, 60, 97

M. Le Muzic, L. Autin, J. Parulek, and I. Viola. cellVIEW: A tool for illustrative and multi-
scale rendering of large biomolecular datasets. In *Eurographics Workshop on Visual Computing
for Biology and Medicine, VCBM 2015*, pages 61–70, Chester, UK, September 14–15, 2015.
DOI: 10.2312/vcbm.20151209. 3

E. H. Lee, J. Hsin, M. Sotomayor, G. Comellas, and K. Schulten. Discovery through the computational microscope. *Structure*, 17(10), pages 1295–1306, 2009. ISSN 0969-2126. http://ww
w.cell.com/article/S0969212609003323/abstract. DOI: 10.1016/j.str.2009.09.001. 2

J. Li. AtomEye: An efficient atomistic configuration viewer. *Modelling and Simulation in Materials Science and Engineering*, 11(2), pages 173–177, mar 2003. ISSN 0965-0393. DOI: 10.1088/0965-0393/11/2/305. 7

N. Lindow, D. Baum, and H.-C. Hege. Interactive rendering of materials and biological structures on atomic and nanoscopic scale. *Computer Graphics Forum*, 31(3), pages 1325–1334, 2012. ISSN 1467-8659. DOI: 10.1111/j.1467-8659.2012.03128.x. 49, 50, 51

C. Loop and J. Blinn. Real-time GPU rendering of piecewise algebraic surfaces. *ACM Trans. Graph.*, 25(3), pages 664–670, July 2006. ISSN 0730-0301. DOI: 10.1145/1141911.1141939. 9

W. E. Lorensen and H. E. Cline. Marching cubes: A high resolution 3d surface construction algorithm. In *ACM SIGGRAPH Computer Graphics and Interactive Techniques*, volume 21, pages 163–169, 1987. DOI: 10.1145/37402.37422. 94

J. MacQueen. Some methods for classification and analysis of multivariate observations. In *Proc. of the 5th Berkeley Symposium on Mathematical Statistics and Probability, Volume 1: Statistics*, pages 281–297. University of California Press, 1967. 97

N. Max. Computer representation of molecular surfaces. *IEEE Computer Graphics and Applications*, 3(5), pages 21–29, 1983. ISSN 0272-1716. DOI: 10.1109/mcg.1983.263183. 7

D. Meagher. Octree Encoding: A New Technique for the Representation, Manipulation and Display of Arbitrary 3-D Objects by Computer. Technical Report IPL-TR-80-111, Rensselaer Polytechnic Institute, 1980. 66

T. Möller and B. Trumbore. Fast, minimum storage ray-triangle intersection. *Journal of Graphics Tools*, 2(1), pages 21–28, 1997. DOI: 10.1080/10867651.1997.10487468. 58

J. J. Monaghan. Smoothed particle hydrodynamics and its diverse applications. *Annual Review of Fluid Mechanics*, 44, pages 323–346, 2012. DOI: 10.1146/annurev-fluid-120710-101220. 65

C. Müller, S. Grottel, and T. Ertl. Image-space GPU metaballs for time-dependent particle data sets. In *Proc. of Vision, Modelling and Visualization (VMV '07)*, pages 31–40, 2007. 90

E. F. Pettersen, T. D. Goddard, C. C. Huang, G. S. Couch, D. M. Greenblatt, E. C. Meng, and T. E. Ferrin. UCSF Chimera—A visualization system for exploratory research and analysis. *Journal of Computational Chemistry*, 25(13), pages 1605–1612, 2004. DOI: 10.1002/jcc.20084. 7

H. Pfister, M. Zwicker, J. Van Baar, and M. Gross. Surfels: Surface elements as rendering primitives. In *SIGGRAPH*, pages 335–342, 2000. DOI: 10.1145/344779.344936. 5

W. T. Reeves. Particle systems—a technique for modeling a class of fuzzy objects. *ACM Transactions on Graphics (TOG)*, 2(2), pages 91–108, 1983. DOI: 10.1145/357318.357320. 5

F. Reichl, M. G. Chajdas, J. Schneider, and R. Westermann. Interactive rendering of giga-particle fluid simulations. In *Proc. of High Performance Graphics 2014*, pages 105–116, 2014. 2, 6

G. Reina. *Visualization of Uncorrelated Point Data*. Ph.D. thesis, Visualisierungsinstitut der Universität Stuttgart, 2008. http://elib.uni-stuttgart.de/opus/volltexte/2009/4517. 20

G. Reina and T. Ertl. Hardware-accelerated glyphs for mono—and dipoles in molecular dynamics visualization. In *Eurographics/IEEE VGTC Symposium on Visualization*, pages 177–182, 2005. DOI: 10.2312/VisSym/EuroVis05/177-182. 9, 16

G. Reina, K. Bidmon, F. Enders, P. Hastreiter, and T. Ertl. GPU-based hyperstreamlines for diffusion tensor imaging. In *Proc. of the 8th Joint Eurographics/IEEE VGTC Conference on Visualization*, EUROVIS'06, pages 35–42, Aire-la-Ville, Switzerland, Switzerland, 2006. Eurographics Association. ISBN 3-905673-31-2. http://dx.doi.org/10.2312/VisSym/EuroVis06/035-042. DOI: 10.2312/VisSym/EuroVis06/035-042. 9

F. M. Richards. Areas, volumes, packing, and protein structure. *Annual Review of Biophysics and Bioengineering*, 6(1), pages 151–176, 1977. DOI: 10.1146/annurev.bb.06.060177.001055. 90

J. S. Richardson. The anatomy and taxonomy of protein structure. *Advances in Protein Chemistry*, 34, pages 167–339, 1981. DOI: 10.1016/s0065-3233(08)60520-3. 84, 89

T. Schafhitzel, E. Tejada, D. Weiskopf, and T. Ertl. Point-based stream surfaces and path surfaces. In *Graphics Interface 2007*, GI '07, pages 289–296, New York, NY. ACM, 2007. ISBN 978-1-56881-337-0. http://doi.acm.org/10.1145/1268517.1268564. DOI: 10.1145/1268517.1268564. 5

K. Scharnowski, M. Krone, F. Sadlo, P. Beck, J. Roth, H.-R. Trebin, and T. Ertl. 2012 IEEE visualization contest winner: visualizing polarization domains in barium titanate. *IEEE Computer Graphics and Applications*, 33(5), pages 9–17, 2013. ISSN 0272-1716. http://scaviscontest.ieeevis.org/. DOI: 10.1109/mcg.2013.68. 90

Schrödinger, LLC. The PyMOL Molecular Graphics System, Version 1.8. PyMOL, The PyMOL Molecular Graphics System, Version 1.8, Schrödinger, LLC., November 2015. 7

C. Sigg, T. Weyrich, M. Botsch, and M. Gross. GPU-based ray-casting of quadratic surfaces. In *Proc. of the 3rd Eurographics/IEEE VGTC Conference on Point-Based Graphics*, SPBG'06, pages 59–65, Aire-la-Ville, Switzerland, Switzerland, 2006a. Eurographics Association. ISBN 3-905673-32-0. http://dx.doi.org/10.2312/SPBG/SPBG06/059-065. DOI: 10.2312/SPBG/SPBG06/059-065. 20

C. Sigg, T. Weyrich, M. Botsch, and M. Gross. GPU-based ray-casting of quadratic surfaces. In *EG/IEEE VGTC Conference on Point-Based Graphics*, pages 59–65, 2006b. DOI: 10.2312/SPBG/SPBG06/059-065. 9

S. W. Skillman, M. S. Warren, M. J. Turk, R. H. Wechsler, D. E. Holz, and P. M. Sutter. Dark Sky Simulations. Early data release, 2014. 3

J. Staib, S. Grottel, and S. Gumhold. Visualization of particle-based data with transparency and ambient occlusion. *Computer Graphics Forum*, 34(3), pages 151–160, 2015. DOI: 10.1111/cgf.12627. 78, 80, 81

M. Tarini, P. Cignoni, and C. Montani. Ambient occlusion and edge cueing for enhancing real time molecular visualization. *IEEE Transactions on Visualization and Computer Graphics*, 12(5), pages 1237–1244, 2006. DOI: 10.1109/tvcg.2006.115. 74, 81

R. Toledo and B. Lévy. Extending the graphic pipeline with new GPU-accelerated primitives. Technical report, INRIA-ALICE, 2004. 9

I. Wald, A. Knoll, G. P. Johnson, W. Usher, V. Pascucci, and M. E. Papka. Cpu ray tracing large particle data with balanced p-k-d trees. In *Proc. of the 2015 IEEE Scientific Visualization Conference (SciVis)*, SCIVIS '15, pages 57–64. IEEE Computer Society, 2015. ISBN 978-1-4673-9785-8. DOI: 10.1109/scivis.2015.7429492. 2

Y.-F. Wang, R. Dutzler, P. J. Rizkallah, J. P. Rosenbusch, and T. Schirmer. Channel specificity: structural basis for sugar discrimination and differential flux rates in maltoporin1. *Journal of Molecular Biology*, 272(1), pages 56–63, 1997. DOI: 10.1006/jmbi.1997.1224. 75

J. Yu and G. Turk. Reconstructing surfaces of particle-based fluids using anisotropic kernels. *ACM Transactions on Graphics*, 32(1), pages 5:1–5:12, 2013. ISSN 0730-0301. http://doi.acm.org/10.1145/2421636.2421641. DOI: 10.1145/2421636.2421641. 90

Y. Zhang, B. Solenthaler, and R. Pajarola. Adaptive sampling and rendering of fluids on the GPU. In *IEEE/ EG Symposium on Volume and Point-Based Graphics*, pages 137–146, 2008. ISBN 978-3-905674-12-5. DOI: 10.2312/VG/VG-PBG08/137-146. 6

Y. Zhu and R. Bridson. Animating sand as a fluid. *ACM Transactions on Graphics*, 24(3), pages 965–972, 2005. ISSN 0730-0301. http://doi.acm.org/10.1145/1073204.1073298. DOI: 10.1145/1073204.1073298. 90

S. Zhukov, A. Iones, and G. Kronin. An ambient light illumination model. In *Eurographics Workshop on Rendering*, pages 45–56, 1998. DOI: 10.1007/978-3-7091-6453-2_5. 74, 76, 77

Authors' Biographies

MARTIN FALK

Martin Falk is an assistant lecturer in the Scientific Visualization Group at Linköping University. He received his Ph.D. degree (Dr. rer. nat.) from the University of Stuttgart in 2013. His research interests are volume rendering, visualizations in the context of systems biology, large spatio-temporal data, glyph-based rendering, and GPU-based simulations. He is the main developer and software architect of the visualization tool CellVis [Falk et al., 2013b] for particle-based data in systems biology.

SEBASTIAN GROTTEL

Sebastian Grottel received his Ph.D. in computer science (Dr. rer. nat.) at the University of Stuttgart, Germany, at the Visualization Research Center (VISUS). Since 2012, he is a post-doctoral researcher at the Chair for Computer Graphics and Visualization at the TU Dresden, Germany. His research interests include high performance graphics for scientific visualization of dynamic data. He is the main developer and software architect of the visualization framework MegaMol [Grottel et al., 2015] for particle-based data.

MICHAEL KRONE

Michael Krone is a postdoctoral researcher at the Visualization Research Center of the University of Stuttgart (VISUS). He received his Ph.D. computer science (Dr. rer. nat.) in 2015 from the University of Stuttgart, Germany. His main research interest lies in biomolecular visualization and visual analysis for structural biology including molecular graphics, particle-based rendering, and GPU-accelerated computing.

GUIDO REINA

Guido Reina is a postdoctoral researcher at the Visualization Research Center of the University of Stuttgart (VISUS). He received his Ph.D. in computer science (Dr. rer. nat.) in 2008 from the University of Stuttgart, Germany. His research interests include large displays, particle-based rendering, and GPU-based methods in general. He is a principal investigator of the subproject *Visualization of Systems with Large Numbers of Particles* for the Collaborative Research Center (SFB) 716.

Printed in the United States
by Baker & Taylor Publisher Services